ADVANCES IN BIOREMEDIATION AND PHYTOREMEDIATION

Edited by **Naofumi Shiomi**

Advances in Bioremediation and Phytoremediation
http://dx.doi.org/10.5772/67970
Edited by Naofumi Shiomi

Contributors

Sevinç Adiloğlu, Ismael Acosta, Juan Fernando Cárdenas González, Adriana Sarai Rodríguez Pérez, Chunming Su, Wenjie Sun, Sha Zhang, Rosa María Martínez-Espinosa, Sonia Aracil-Gisbert, Javier Torregrosa-Crespo, Guadalupe Virginia Nevárez-Moorillón, Hilda Amelia Piñon-Castillo, Daniel Lardizabal-Gutiérrez, Daniel Hernández-Castillo, Laila N. Muñoz-Castellanos, Blanca E. Rivera-Chavira, Francisco Javier Zavala-Díaz De La Serna, Khaled Abdulaziz Alakeel, Ikrema Hassan, Eltayeb Mohamedelhassan, Ernest Yanful, Ze-Chun Yuan, Feyisayo Victoria Adams, Maryam F. Awode, Bolade Agboola, Zeljka Rudic, Bojana Vujovic, Ljubinko Jovanovic, Igor Kljujev, Mile Bozic, Vera Raicevic, Dragan Kikovic, Naofumi Shiomi

Published by InTech
Janeza Trdine 9, 51000 Rijeka, Croatia

Publishing Process Manager Edi Lipovic
Technical Editor SPi Global
Cover InTech Design team

First published April, 2018
Printed in Croatia
Legal deposit, Croatia: National and University Library in Zagreb

Additional hard copies can be obtained from orders@intechopen.com

Advances in Bioremediation and Phytoremediation, Edited by Naofumi Shiomi
p. cm.
Print ISBN 978-953-51-3957-7
Online ISBN 978-953-51-3958-4

For EU product safety concerns:
IN TECH d.o.o., Prolaz Marije Krucifikse Kozulić 3, 51000 Rijeka, Croatia,
info@intechopen.com or visit our website at intechopen.com.

Meet the editor

Dr. Naofumi Shiomi studied on the recombinant yeast as a researcher at the Laboratory of Production Technology of Kanena Corporation for 15 years until 1998 and earned his PhD degree in Engineering from Kyoto University. He now works as a professor at the School of Human Sciences of Kobe College in Japan, where he teaches biotechnology and life science in his "Applied Life Science" laboratory. He has studied bioremediation and biomedical science for 21 years at Kobe College and has published more than 40 papers and several books. His recent research has focused on the prevention of obesity and aging.

Contents

Preface

Human beings have taken good care of the use of freshwater resources, which only occupy 0.01% of the planet's total water, and have in turn reaped the benefits. However, the limited water resources have reached a crisis state. The increased population size has resulted in the necessity of increased production of food and chemical compounds that use huge amounts of water, and this has consequently caused a shortage of drinking water. In addition, living and industrial drainages, emission gasses from industries, and the disposal of unnecessary ore containing heavy metals have increased and are being poured into rivers or exhausted into the air without adequate pretreatment. Those pollutants often result in the fatal condition of rivers and groundwater. Currently, agricultural land where crops can no longer grow and groundwater that can no longer be used as drinking water are increasing at surprisingly high speed.

The reports published by the United Nations in 2007 stated that around 1.1 billion people live with a serious shortage of water and 1.8 million children die every year from dirty drinking water because of the development of various diseases. Moreover, the number of countries in which a shortage of water is reaching an extremely serious level is approximately 30 to 40, which means that 660 million people find it difficult to obtain water sources. Such areas will be expanded much more in the near future, and it is assumed that over 60% of people all over the world will suffer from a serious shortage of water by 2025.

The emission of harmful compounds must be banned by strict laws or treaties as soon as possible. Several directives and international treaties have already been promulgated. However, compared to the EU, they do not function sufficiently in many countries in Asia, South America, and Africa, and the areas of serious pollution are unfortunately expanding in these regions. If soil and groundwater are completely polluted, it will be difficult to remediate the situation. Therefore, large-scale projects for the remediation of polluted areas must be performed to avoid a fatal condition. However, the combination of bioremediation processes that are applied to narrow areas is insufficient for the remediation of a large area. Therefore, the development of novel high-performance bioremediation processes is desired.

Thus, this book deals with special strains and a combination of various processes that would make bioremediation and phytoremediation processes more prominent. In the chapters of this book, the researchers have introduced the recent status of pollution; the characteristics of bioremediation using halobacteria, *Candida* yeast, and autochthonous bacteria; and phytoremediation using plants and macrophytes. Moreover, other researchers have introduced remediation processes using biochar and electric currents. These chapters will provide readers with new ideas on effective bioremediation.

I edited similar topics in the book *Advances in Bioremediation of Wastewater and Polluted Soil* (InTechOpen book in 2015), and it complements the purpose of this book. Therefore, I recommend that interested readers consult it in addition to this book. Finally, I would like to thank Mr. Edi Lipović and the publishing managers of InTechOpen Publishers for their support throughout the publication process.

Naofumi Shiomi
Kobe College
Japan

Chapter 1

Introductory Chapter: Serious Pollution of Soil and Groundwater and the Necessity of Bioremediation

Naofumi Shiomi

Additional information is available at the end of the chapter

http://dx.doi.org/10.5772/intechopen.74403

1. Introduction

Earth has massive amounts of water resources, but drinking water accounts for only 0.01% of the total water. Human beings have taken good care of the use of our precious water resources and had benefitted sufficiently. The limited water resources, however, have reached a critical state. A report published by the United Nations in 2007 suggested that around 1.1 billion people live with serious water shortages and 1.8 million children die every year because they develop various diseases from drinking dirty water [1, 2]. To make matters worse, water shortages have already reached extreme levels for 660 million people living in approximately 30–40 countries. Moreover, the area where drinking water cannot be sufficiently obtained is constantly increasing.

It is estimated that over 60% of people will suffer serious water shortages by 2025. A serious shortage of food resources also occurs because crops cannot be grown in countries where there are water shortages or where water is polluted. The prospect of the eruption of international disputes, or even war, over water sources is also worrying. If the emission of harmful pollutants is not stopped and effective remediation is not performed as soon as possible, the global environment will fall into a terrible situation. As this is the introductory chapter of this book, the author introduces the most recent conditions of pollutants and the causes of pollution. Moreover, the author suggests the ideal bioremediation and phytoremediation processes to apply to large polluted areas.

2. Causes of the shortage of drinking water

One of the causes of the shortage of drinking water is the excessive use of river water. An increase in population caused an increase in industrial activities and food production, and many large-scale projects that use irrigation water from rivers for agricultural land have been implemented. Moreover, a massive amount of domestic and industrial water has been used by an increasing population and as a result of the rapid development of world economies. The excessive use of water has caused a shortage of river water. In countries located at or near downstream basins, the shortage is especially serious because river water is excessively consumed by the countries located upstream. For example, Egypt is a country where the desert occupies 95% of the land and where the water of the Nile River is the sole water source. Since the countries located in the upstream area of the Nile River excessively consume water and pollutes the river, Egypt is now troubled by serious water shortages [3]. Similar problems have occurred in the countries located at the Indus river basin. In addition to these causes, increases in population and industrial activities have increased the amount of greenhouse gasses that cause global warming, and abnormal weather such as El Nino events occurs more frequently now. Droughts and flooding as a result of abnormal weather make it difficult to stock drinking water.

The deterioration of water quality as well as the amount is another cause of the shortage of drinking water. Living drainage and industrial wastewater that does not receive proper pre-treatment are increasing along with the increase in population. This wastewater is directly poured into rivers and the water quality decreases to inadequate levels and can no longer be used as drinking water [4, 5]. Moreover, excessively used herbicides and exhaust gasses containing various chemical compounds and heavy metals contaminates the soil and groundwater. In countries where the economy is rapidly developing, the deterioration of water quality from these causes is especially serious. Many farm lands are polluted by excessively used fertilizer and herbicides or heavy metals and can no longer be used. Fluorine and arsenic have contaminated the water in tube wells due to soil pollution, and inhabitants that drink from these wells often develop serious cases of poisoning [6, 7]. The typical pollutants are listed in **Table 1**, and the author has introduced each cause more precisely in the next section.

Main poisoning sources	Main chemical compounds	Serious contaminated sites	Remarks on risks to the environment
Acid rain	SOx and NOx	Urban places and thermal power stations	Death of fish and aquatic organisms in lakes and acid shock in forests
Insecticides	POPs (BHC, DDT)	Agricultural land	Contamination of crops and drinking water, or direct absorption from polluted air
	OPPs (fenitrothion, diazinon)	Agricultural land	
Herbicides	Atrazine, simazine		
Fertilizer	Nitrate nitrogen	Agricultural land	Eutrophication of soil and rivers due to excess nitrate nitrogen
Metalloids	As, F	Urban, mining, and metallurgy sites	Contamination of tube wells
Heavy metals	Pb, Cd, Hg, Cr (IV). etc.		Contamination of crops and drinking water or direct absorption from polluted air

Table 1. Pollution that occurs in widespread areas in developing countries.

3. Pollutants causing serious damage

3.1. Acid rain

One of the pollutants causing serious environmental pollution is acid rain. Sulfur and nitrogen oxides (SOx and NOx) are excessively exhausted from factories and cars and acidify rain. In the USA and European countries, in the 1970s, serious acidification of soil and lakes by acid rain occurred in association with economic development. In those days, acid rain at pH 2–3 was recorded in several areas in the United States, and the harmful effects on the ecosystem were observed in the acidification of lakes in European countries. Mg^{+} and Ca^{2+} ions, which are important minerals to build bones, were leached from lakes by acidification. Fish are most sensitive to the loss of those ions, and many species of fish with bending bones or abnormal structures appeared in acidified lakes including Little Rock Lake in Sweden [8]. The investigation conducted on 3821 lakes in Norway, Sweden, and Finland suggested that many species of fish living in those lakes had disappeared due to acidification [9].

Acid rain induced the acidification of soil as well as lakes. Soil particles charge negative under normal conditions and can maintain a sufficient amount of minerals for the growth of plants. However, the acidification of soil promotes the leaching of minerals such as Mg^{2+} and K^{+} from the soil and the accumulation of harmful metal ions in the soil. Furthermore, the acidification of the soil decreases soil microorganisms that degrade sediments and supply the soil with nutrients, and it also decreases microorganisms that assist with the absorption of nutrients and the adjustment to normal pH values. With the decrease of minerals, microorganisms in the soil can finally cause serious damage to plants. Acid shock, the phenomena where all trees in a forest suddenly wither, had unfortunately occurred in several forests in North America and Europe [10, 11].

Once aquatic organisms and fish have died or forests have been lost to acid shock, it takes several decades for it to return to its original state. Therefore, remediation by neutralization is necessary before serious damage occurs. The neutralization of acid with calcium carbonate and calcium hydrate powder is a general and inexpensive procedure [12, 13]. Several projects to neutralize lake water were conducted by spraying the powder from a small ship or helicopter. For instance, 3 million dollars were invested in the neutralization of lakes with calcium carbonate powder in Sweden. Recently, the acidification of lakes and soil has considerably improved in the European Union (EU) and North America due to the efforts to decrease the emission of SOx and NOx and the implementation of projects to neutralize lakes.

On the other hand, the problem of acid rain has gradually occurred in countries in Asia, Africa, and South America. For example, Asian countries including China, India, and Vietnam have realized remarkable economic development during the last half century. The development of industry and an increase in the demand for cars and electronics has caused the release of more emissions containing SOx and NOx. Huge amounts of dust containing PM 2.5 and photochemical smog often appear in urban areas in China and India [14]. As acid rain continuously falls in those areas and effluent control of SOx and NOx by treaties or directives does not function sufficiently, it is difficult to stop or decrease those emissions. Therefore, a serious condition that is similar to that of Europe and the USA in the 1970s will occur in the near future in those countries. The neutralization of soil and lakes by using microorganisms [15, 16] may be necessary until emissions can be controlled successfully.

3.2. Agricultural chemicals

Pesticides are often used excessively because they generally enhance the productivity of crops. Excess use, however, has harmful effects on the soil. Persistent organic pollutants (POPs) such as dichlorodiphenyltrichloroetane (DDT) and benzene hexachliride (BHC) had been used most frequently as an insecticide in the 1980s. POPs were stable and stayed in the soil for long periods of time [17] and excess use caused the contamination of groundwater and plants. Several studies reported that birds and fish accumulated POPs in their bodies, and milk from cows that ate contaminated straw contained POPs. Thus, the production and use of POPs were banned at the Stockholm Convention on Persistent Organic Pollutants held in 2001, except for its use to kill mosquitoes carrying malaria. Organophosphorus pesticides (OPPs) such as parathion and diazinon were developed as substitutes for POPs. OPPs were thought to be safe because the length of time that they remain in soil was much shorter than that of POPs. However, contamination of drinking water by OPPs was reported in the European Union (EU) in the 2000s. As OPPs inhibit acetylcholine receptors in nervous systems, drinking water containing OPPs may have toxic effects on the nervous system [18]. Thus, the EU banned the use of OPPs, although they are still used in other countries.

The excessive use of herbicides also causes serious soil pollution. The herbicides that are used most frequently are triazine herbicides such as atrazine and simazine which were developed in the 1970s, and have been used in Europe and the United States. However, frogs with deformed fingers were found in a paddy field where atrazine had been used at a high concentration, and it was clarified that they functioned as an endocrine disrupter. Other researchers warned that the decrease of frogs all over the world may be caused by the excessive use of atrazine. Based on these reports, the EU decided on an upper limit for the use of atrazine in fields and the Environment Protection Agency (EPA) in the United States legally regulated the maximum amount (3 ppm) of atrazine in drinking water. However, atrazine and simazine are still widely used outside of the EU and polluted paddy fields are increasing [19, 20].

In order to decrease the harmful effects, insecticides and herbicides were exchanged for low persistence products. Pyrethroid analogs (chrysanthemic acid and pyrethorolone) and nicotine analogs (imidacloprid) were developed to decrease the persistence and toxicity, and they are widely used. Moreover, studies that aimed to decrease the amount of herbicides were also conducted. For example, the Monsanto company developed genetically modified organisms (GMOs) that were resistant to Roundup, which was an herbicide containing glyphosate isopropane as its main component. The amount of Roundup used could be decreased by the use of the GMO, although the decreased amount was much less than expected. Other groups developed insecticides that could attach to leaves more tightly to decrease the necessary amount. Nevertheless, in spite of these efforts to decrease the amount of insecticides and herbicides, farm fields where crops cannot be grown anymore are rapidly increasing, and this is becoming a serious problem in many countries especially in Asia, Africa, and South America

[21–23]. To make matters worse, a huge amount of POP stock produced before the Stockholm Convention on Persistent Organic Pollutants still exist, and POPs are still used in several areas in Asia. According to the assessment of FAO/UNEP/WHO, 1–5 million people have received health damage, and several thousand die every year because of soil pollution [24].

3.3. Harmful metals

Harmful metals are also causing serious pollution on a global scale. For instance, arsenic (As) is found in arsenopyrite, and sulfide minerals composed of ferric and arsenic are widely present in soil. Concentrations of As are 2–23 μg/g in soil and 0.005–0.1 $\mu g/m^3$ in the air. Until the 1980s, products using As were not frequently found and those concentrations were sufficiently low and adhered to safety levels. However, As production rapidly increased because Ga-As and Se-As semiconductors were widely used in cellphones, personal computers, and other appliances, and the mining industry that had to produce As were rapidly increasing accompanied by the increased demand. As a result, huge amounts of crude ore, sediments, wastewater, and emission gasses containing As contaminated rivers, groundwater, and soil [25, 26]. Serious health damage has occurred in inhabitants living in areas that surround mining industries.

Furthermore, high concentrations of As are observed in groundwater in several Asian countries including Vietnam, Thailand, India, and China [27–29]. Many people in those countries use water in tube wells as drinking water, but high concentrations of As have contaminated these wells. In the guideline of the World Health Organization (WHO), the concentration limit of As to avoid harmful effects is 10 μg/L, but the groundwater in those tube wells often exceed 50 μg/L. India and Bangladesh are especially troubled by the contamination. Serious groundwater pollution by As in West Bengal prefecture has been reported and approximately 8 million inhabitants in the prefecture are exposed to the risk of arsenic poisoning. The person who noticed arsenic poisoning as a result of drinking groundwater from tube wells in the prefecture had not reported it until 1983, and the data suggested that the concentration of As in groundwater in the tube wells has been gradually increasing after 1983. Arsenopyrite may be melted by the acidification of soil and release As, although the cause of the increase has not been sufficiently elucidated.

Coal fuel is the third most important cause of As pollution. The content of As in coal fuel used at Guizhou in China is 100–9600 ppm; however, the As content in coal fuel is usually 1–10 ppm, and therefore, the inhabitants have developed serious As poisoning. The Chinese government has now banned the mining of such low-quality coal. Demand for coal fuel is increasing and is accompanied by the development of economies in Asia. Poisoning by As is also increasing in countries where coal is the main fuel source. Furthermore, coal fuel causes fluorine poisoning as well as As poisoning. In some areas in Sichuan and Guizhou in China, coal containing very high levels of fluorine (500 mg/kg) is used, and the inhabitants have developed fluorine poisoning. According to the report of the Chinese Ministry of Health in 1997, around 20 million people developed fluorosis caused by the use of coal as a fuel source. Moreover, many people who drank groundwater from tube wells in India also developed fluorosis in addition to arsenic poisoning.

Other forms of health damage are caused by pollution from harmful heavy metals such as lead, cadmium, mercury, and hexavalent chromium [30–34]. For instance, contamination by lead and cadmium has recently become serious. Lead poisoning has historically occurred through the organic lead exhausted from cars using lead gasoline (gasoline containing tetraethyl or tetramethyllead) in the 1970s, but from 1980 to 1990, lead gasoline was banned in many countries and changed to lead-free gasoline. As a result of this policy, the lead concentration in the air decreased to the normal level, and the problem was solved. However, lead poisoning became a problem once again. More than 80% of the lead currently utilized is for lead-acid batteries which are used as car batteries. The production amount of lead-acid batteries is rapidly increasing due to the increased demand for cars. In many metallurgy and mining industries, exhaust gasses and wastewater containing high concentrations of lead are directly exhausted without any pretreatment or with insufficient pretreatment, and low-quality ore and used sludge are left on the soil [35, 36]. Therefore, serious lead poisoning has developed in areas surrounding metallurgy and mining industries. The cause of contamination by cadmium is similar to that of lead. Cadmium is mainly used in Li-Cd batteries for home appliances, and the demand is rapidly increasing. As the recycling ratio of Li-Cd batteries is low (around 20%), cadmium poisoning has been reported in various areas as well as in those surrounding mining industries. Burnt ash and incombustible garbage including waste from electrical and electronic equipment (WEEE), metal plates, and pipes also contain heavy metals [37]. The amount of WEEE is steadily increasing at a rate of 5%, and it was approximately 9 tons in 2005 in the EU [38].

Unfortunately, the Japanese have experienced several tragedies resulting from heavy metal poisoning. A Japanese mining company eliminated cadmium waste into nearby rivers and the inhabitants developed "ouch-ouch disease," which exhibited symptoms such as dizziness, leg pain, and fragile bones. In another case, a company eliminated waste containing methylmercury into a nearby river and it resulted in a form of neuroparalysis named "Minamata disease." Recently reported metal contamination of groundwater and rivers in China and India is much worse than the case in Japan. Therefore, we have to pay attention to this problem and perform countermeasures as soon as possible.

4. Countermeasures to remediate pollution occurring on a global scale

4.1. Best countermeasures for eliminating pollution from soil and groundwater

Contamination of soil and groundwater has worsened in countries in Asia, South America, and Africa as described in Section 3, and therefore, those countries have to take effective countermeasures without delay. The best countermeasure to stop the progress of such contamination is to decrease the emission of pollutants with strict legal regulation and treaties.

We can remark on the case of acid rain as a successful model case. In order to improve the problem of acid rain, the Canada-U.S. Air Quality Agreement had decided on the restriction of SOx emissions and acted accordingly. As a result, the pH of lakes increased gradually. In the lakes of Killarney Park, the pH value of the lake water increased to nearly 6.0 with the decrease of air pollutants in Sudbury [39], and the acidification of soil in forests such as Boemina Forest and lakes in Europe such as Finishing lake was gradually improved as a result of the decrease in air pollution. Moreover, the convention on Long-range Transboundary Air Pollution (LRTAP) which was attended by 48 countries was conducted in 1983, and the emission of polluted air was more strictly controlled. These efforts had good results in the decrease of damage done by acid rain [40].

We can also remark on the directives of WEEE in the EU as another successful example. The EU has created several countermeasures to solve the problems of increased garbage containing harmful metals. First, the EU promulgated the Directive of End-of-Life Vehicles (ELV) in 2000, which necessitated the recycling of ELV to prevent the illegal dumping of used batteries. Second, the EU promulgated two directives, the WEEE Directives and the RoHS Directive, in 2003. Recycling of WEEE was obligated, and the amount of harmful metals used in WEEE was strictly restricted by those directives. The target products of the RoHS Directives were expanded to approximately 20,000 products by the revised RoHS Directive (RoHS2) in 2011. Countries exporting electric appliances to the EU had to follow these strict directives. As a result of these strict directives, the ratio of recycling of lead-batteries was over 90%, and substitutes without harmful metals such as alkaline batteries, lead-free solders, and lead-free glass have been developed.

In spite of the countermeasures in the EU, 44.7 million tons of WEEE were discarded globally in 2016 and, according to a report by Kokuren University in Japan, this amounts to an increase of 8% from 2013. The total ratio of the recycling of WEEE is only 20%. The result suggests that further efforts to decrease WEEE are necessary in other countries as well as in the EU. Similar directives to that of the EU are promulgated in several countries in Asia. However, it is difficult to keep pace among Asian countries where economies have rapidly developed and emission control is insufficient because the development of the economy is more important than environmental problems. If strict regulation by legal countermeasures is not conducted as soon as possible, there is no doubt that soil and groundwater will fall into a dilapidated state in the near future.

4.2. Best process to remediate the serious pollution of soil and groundwater

Once groundwater and soil have been fatally polluted, several decades are needed for it to recover even if the emission of pollutants can be stopped by strict directives and treaties. Therefore, the remediation of soil and groundwater must be implemented in parallel with efforts to decrease emissions. Various remediation processes such as bioventing, bioremediation with Fenton reactions, and oxygen release compounds (ORC) have been developed, and those conventional processes are successfully applied to contamination spread in a narrow area [41, 42]. However, they are inadequate when pollution has spread in a large area because they are expensive processes. Novel and inexpensive bioremediation processes that can be applied to pollution spread on a global scale must be developed.

An outline of the ideal bioremediation process for contamination of a massive area, which is recommended by the author, is shown in **Figure 1**. In order to realize such a remediation process, the zone where remediation is performed is an important factor. Emission gasses containing various harmful pollutants contaminate the air, fall on the soil as rain, and penetrate the soil surface. This process first causes the accumulation of pollutants in the shallow layer of the soil. Following this process, the polluted zone is gradually expanded to a deeper zone of the soil, and pollutants finally reach the deepest zone where groundwater is present.

At the deeper and deepest zones, the oxygen concentration and the amount of microorganisms are very low, and large-scale equipment and high pressure pumps that supply oxygen and microorganisms are very expensive and are necessary in the remediation process. Therefore, the remediation of pollutants is extremely difficult once pollutants reach the deeper or deepest zones. On the other hand, remediation at the shallow layer of the soil has many advantages [43]. The addition and control of both nutrients and microorganisms to the soil is very

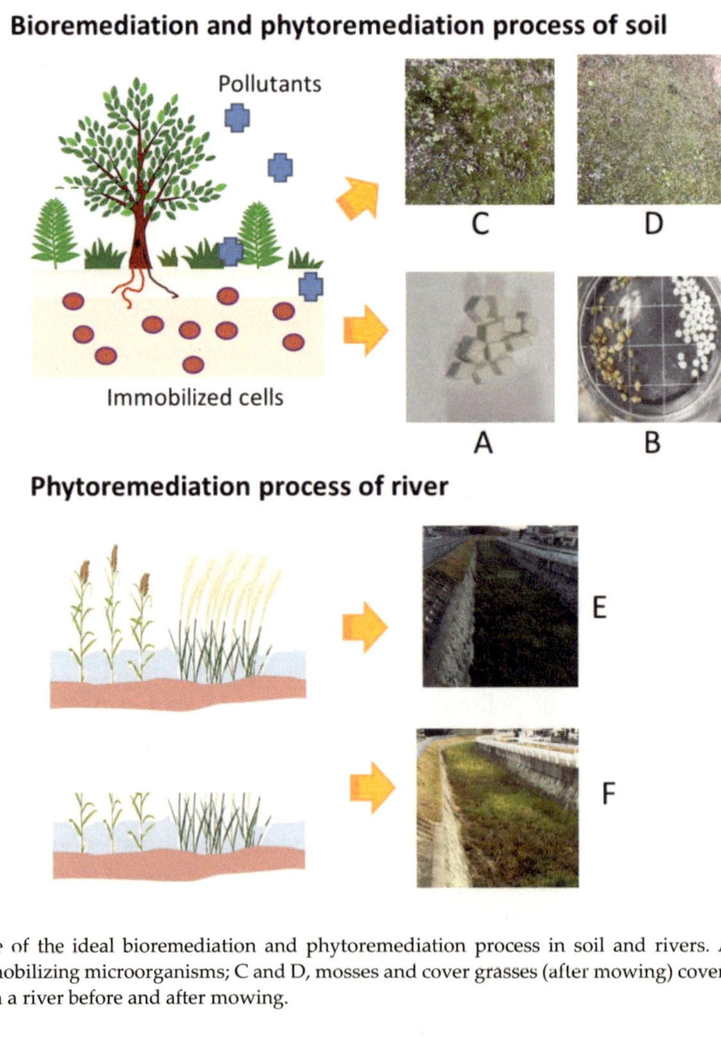

Figure 1. Scheme of the ideal bioremediation and phytoremediation process in soil and rivers. A and B, BSPs and alginate gels immobilizing microorganisms; C and D, mosses and cover grasses (after mowing) covering the soil surface; E and F, plants in a river before and after mowing.

easy at the shallow layer because sufficient oxygen is present in the zone, and nutrients and microorganisms can be supplied by using a normal sprayer. Furthermore, since plant roots that may absorb pollutants cannot reach under several meters of soil, remediation at the shallow layer is absolutely necessary in the case of phytoremediation. These aspects suggest that the removal of the pollutants at the shallow layer of the soil is an absolute requirement in order to realize the ideal remediation process.

As rain transports pollutants to a deeper zone, the retention time of pollutants in the shallow layer of the soil is not long. Therefore, a device is necessary to implement the rapid removal of pollutants within the shallow layer. To realize this, microorganisms showing abilities of degradation and absorption of pollutants must be immobilized at high densities in the shallow layer. Several kinds of immobilization technology can be used for this purpose, and immobilization with alginate gel or κ-carrageenan gel is the general procedure. The advantage of this method is that it can be applied to various kinds of nonflocculent microorganisms. Otherwise, in the case where the microorganisms show characteristics of flocculent or fungi-like shapes, biomass-supported particles (BSPs) (photo A in **Figure 1**) and self-immobilization methods (BSIS) are more useful and inexpensive processes [44–46]. For instance, when *Bacillus subtilis* cells, which secrete a viscous polymer, were self-immobilized by just spraying them onto the soil, the cells were immobilized at high densities and the rapid degradation of herbicides could be realized within the shallow layer. This method can be applied to massive areas because of the simple protocol and inexpensive apparatus. Therefore, self-immobilization methods may be best for the degradation of chemical compounds. However, in the event of the removal of harmful heavy metals, microorganisms must be collected from the soil after the absorption of metals. In such a case, immobilization with alginate gels is more prominent than the self-immobilization method because microorganisms immobilized with alginate gel can be easily collected from the soil using a sieve as shown in photo B in **Figure 1** [30].

If absorption processes can be implemented by covering the surface of the soil as well as at the shallow layer, more prominent remediation processes can be constructed. Mosses may be applied to this process as shown in photo C in **Figure 1**. The advantages of mosses are: (1) they can cover the ground completely without disturbing the flow of rain water, (2) they can grow at a faster rate with little water, and (3) they have advantages in the reuse of metals because they can accumulate metals at high concentrations. In fact, *Scopelophia cataractae* and *Funaria hygrometrica* that are known as hyperaccumulators of metals could accumulate copper and lead at very high levels [47, 48].

Many perennial plants can accumulate metals in their stems and leaves and can reproduce them after the removal of stems and leaves. As pea sprouts can accumulate several heavy metals, it may be one of the desirable plants. Mamenae, the young herbs of pea sprouts, can recover in a week after mowing, and therefore, a continuous adsorption process can be constructed. Moreover, there are many wild grasses known as "ground cover" that are shown in photo D in **Figure 1**. They can cover the ground completely and grow even after mowing. Additionally, they are very strong in harsh climates. Therefore, they may be the best plants to use to realize a continuous process. The plants that can accumulate components in the roots at high levels may also be useful. For example, *Allium cepa* produces onions. If the gene coding adsorbent can be expressed in onions, an inexpensive phytoremediation process can be realized.

In the case of river remediation, another concept is necessary. Once domestic and industrial wastewater containing huge amounts of pollutants is directly discharged into rivers, it is extremely difficult to remove the pollutants. Therefore, the establishment of pretreatment equipment is necessary to remove the pollutants. Many countries, however, cannot afford the cost of waste processing. Inexpensive remediation processes for river pollution must be developed before pollutants deal fatal damage to the living things in the rivers and to the inhabitants near rivers. Plants that can live in rivers may provide effective processes, although it is difficult to realize. Photos E and F in **Figure 1** show the scheme of such a process (this river is not being used for phytoremediation, and therefore, this photo is merely an image for the sake of comprehension). The plants shown in the photos can live in rivers where the depth of the river water is 20–30 cm, and can live after being mowed every 6 months and grow again. These plants can live in rivers for many years. Inexpensive pretreatment processes for river pollution may be realized by using this process.

4.3. Best microorganisms and plants to remediate pollution of soil and groundwater

In order to realize remediation processes for pollution spread on a global scale, it is also important to choose microorganisms and plants with the best capacities. Many microorganisms that are exceptional at degrading or removing pollutants have been discovered. For example, fungi that could neutralize acid by secreting basic compound was discovered [45]. Those fungi may effectively neutralize acidified soil and lakes. Additionally, many microorganisms that show increased abilities to degrade triazine pesticides such as atrazine and organophosphorus herbicides such as diazinon have been discovered, along with many microorganisms that can absorb heavy metals [49–52]. Their degradation pathways or mechanisms of absorption were elucidated, and many genes for degradation have already been cloned. Moreover, many plants that can effectively absorb and accumulate metal ions have been discovered and those plants are known as hyperaccumulators [53, 54]. For example, *Rinorea nicolifera*, which was recently discovered in Western Luzon, Philippines, could accumulate unusually high amounts (18,000 ppm) of nickel [55].

However, in the case of the remediation of a large polluted area, the capacities of those microorganisms and hyperaccumulators are insufficient. The author's rough estimate for lead and nickel suggests that the expenditure of plant growth is much higher than the gains of obtained metal, even if the best hyperaccumulators are used under ideal conditions. Therefore, microorganisms and plants that have even more excellent capacities must be developed. There are several advantages to the use of recombinant microorganisms and plants [56–58]. For instance, the capacity of degradation or absorption per cell in recombinant microorganisms can be enhanced several (or several 10) times, and its amount of expression can be freely controlled by using an adequate promoter.

Another advantage in utilizing recombinant microorganisms and plants is that their capacity for degradation or absorption can be maintained under harsh climates. The activity of microorganisms and the growth of plants are very sensitive to harsh climates, and this is a disadvantage of the biological process. Pollution spread over a large area has mainly occurred under severe climates such as acidic, cold, and dry weather, and this lowers the bioremediation efficiency. Therefore, the addition of the capacity to maintain high activity under harsh climates by using gene technology is very important. In the case of cold climates where the activity of microorganisms is inhibited, the use of cold-resistant microorganisms may be adequate as a host strain for gene manipulation. Cold shock proteins may also be useful for the activation of microorganisms [59]. In the case of a shortage of rainfall, moisturizing of the soil is necessary in order to prevent fatal damage to microorganisms and plants. Polymers such as poly-glutamic acid, chondroitin, hyaluronic acid, and those microorganisms or polymers that are secreted from animal cells and microorganisms can be used as soil moisturizing agents [60]. Moreover, several plants can grow in dry climates. For instance, an aloe can grow with little water, accumulate huge amounts of moisture components in its leaves, and continuously grow new leaves without withering. Therefore, aloes have become a prominent host plant for gene manipulation.

The combination of microorganisms and plants as well as the enhancement of their capacity is important in the enhancement of effectiveness. Plants and microorganisms help each other live in natural places [61, 62]. For instance, plants grow by photosynthesis and their fallen leaves or withered plants are degraded by soil microorganisms. The degraded products become nutrients for microorganisms and plants. This energy cycling is necessary to realize the remediation process for long periods of time at a high level of performance. Therefore, the best selection and combination is necessary to realize the energy cycle of the coculture of microorganisms and plants at the polluted site. This is important for the construction of an effective process. In the case of the adsorption of harmful metal ions, synergy is especially important. Many microorganisms show high resistance to heavy metals and high abilities of adsorption, but the removal of microorganisms adsorbing metals from the soil is difficult. On the other hand, although the removal of plants absorbing metals from the soil is easy, resistance to heavy metals in plants is lower than that of microorganisms and they cannot absorb pollutants at the deeper zone (lower than 1–2 m in depth). Therefore, a combination of microorganisms and plants makes the remediation process more effective.

The removal efficiency of metal ions can be enhanced further by using the electric method. Recently, microorganisms that can generate a current have been reported [63–66]. Novel electric processes may be constructed by using those microorganisms, although it is not sufficient to utilize them as a power supply. These aspects suggest that synergetic utilization among plants, microorganisms, and chemical reactions are absolutely necessary to construct the best remediation processes. With the exception of the microorganisms listed earlier, there are many microorganisms that show specific characteristics, and they will become the fighting powers in the remediation process. The author expects that a process that will overcome the spread of serious pollution on a global scale will be developed in the near future.

5. Conclusion

In this introductory chapter, the present conditions of soil and groundwater pollution occurring all over the world, and the necessity of bioremediation have been introduced. Pollution in several Asian, African, and South American countries is much more serious than most people image, and there will be a shortage of food and drinking water in the world in the near future. Strict restrictions by directives or international treaties are the only way to limit and lessen pollution, but the effort is currently insufficient. Therefore, large-scale projects for remediation must be performed in parallel with the effort to stop the emission of pollutants before it results in a fatal condition. Bioremediation and phytoremediation are the most prominent procedures for remediation, but the enhancement of performance is absolutely necessary to fight pollution on a global scale. The author thinks that (1) rapid remediation at the shallow layer of the soil, (2) the use of high-performance recombinant microorganisms that are resistant to harsh climates, and (3) synergy among microorganisms, plants, and chemical reactions that are necessary to realize such a remediation system.

Author details

Naofumi Shiomi

Address all correspondence to: n-shiomi@mail.kobe-c.ac.jp

Kobe College, Japan

References

[1] Reports of United Nations "Water scarcity". http://www.un.org/waterforlifedecade/scarcity.shtml

[2] Reports of UNICEF "Thirsting for a Future: Water and children in a changing climate)". 2017

[3] Omar Mel D, Moussa AM. Water management in Egypt for facing the future challenges. Journal of Advanced Research. 2016;**7**(3):403-412. DOI: 10.1016/j.jare.2016.02.005

[4] Chen WP, Zhang WL, Pan N, Jiao WT. Ecological risks of reclaimed water irrigation: A review. Huan Jing Ke Xue. 2012;**33**(12):4070-3080

[5] Lu S, Wang J, Pei L. Study on the effects of irrigation with reclaimed water on the content and distribution of heavy metals in soil. International Journal of Environmental Research and Public Health 2016;**13**(3). pii: E298. DOI: 10.3390/ijerph13030298

[6] Majumdar KK, Guha Mazumder DN. Effect of drinking arsenic-contaminated water in children. Indian Journal of Public Health. 2012;**56**(3):223-226. DOI: 10.4103/0019-557X.104250

[7] Qiu J. China faces up to groundwater crisis. Nature. 2010;**466**(7304):308. DOI: 10.1038/466308a

[8] Swenson WA, McCormick JK, Simonson TD, Jensen KM, Eaton JG. Experimental acidification of Little Rock Lake (Wisconsin): Fish research approach of early responses. Archives of Environmental Contamination and Toxicology. 1989;**18**:167-174

[9] Tammi J, Appelberg M, Beier U, Hesthagen T, Lappalainen A, Rask M. Fish status survey of Nordic lakes: Effects of acidification, eutrophication and stocking activity on present fish species composition. Ambio. 2003;**32**:98-105. DOI: 10.1579/0044-7447-32.2.98

[10] Evans JS. Biological effects of acidity in precipitation on vegetation: A review. Environmental and Experimental Botany. 1982;**22**(2):155-169. DOI: 10.1016/0098-8472(82)90034-X

[11] Acid Rain: This dangerous precipitation can have serious consequences for the ecosystem. National Gengraphics. https://www.nationalgeographic.com/environment/global-warming/acid-rain

[12] Seoane S, Leirós MC. Acidification-neutralization processes in a lignite mine spoil amended with fly ash or limestone. Journal of Environmental Quality. 2001;**30**(4): 1420-1431

[13] Materechera SA, Mkhabela TS. The effectiveness of lime, chicken manure and leaf litter ash in ameliorating acidity in a soil previously under black wattle (*Acacia mearnsii*) plantation. Bio/Technology. 2002;**85**(1):9-16

[14] Liang G, Liu X, Chen X, Qiu Q, Zhang D, Chu G, Liu J, Liu S, Zhou G. Response of soil respiration to acid rain in forests of different maturity in southern China. PLoS One. 2013;**8**(4):e62207. DOI: 10.1371/journal.pone.0062207

[15] Shiomi N, Katoh S. Bioremediation of soil and water acidified by acid rain with fungi cells. In: Roglesfield EL, editor. Acid Rain Research Focus. USA: Nove Science Publisher; 2011. pp. 105-124

[16] Shiomi N, Yasuda T, Inoue Y, Kusumoto N, Iwasaki S, Katsuda T, Katoh S. Characteristics of neutralization of acids by newly isolated fungal cells. Journal of Bioscience and Bioengineering. 2004;**97**(1):54-58. DOI: 10.1016/S1389-1723(04)70165-6

[17] Li QQ, Loganath A, Chong YS, Tan J, Obbard JP. Persistent organic pollutants and adverse health effects in humans. Journal of Toxicology and Environmental Health A. 2006;**69**(21):1987-2005. DOI: 10.1080/15287390600751447

[18] Eddleston M, Buckley NA, Eyer P, Dawsonb AH. Management of acute organophosphorus pesticide poisoning. Lancet. 2008;**371**(9612):597-607. DOI: 10.1016/S0140-6736(07)61202-1

[19] Litchfield MH. Estimates of acute pesticide poisoning in agricultural workers in less developed countries. Toxicologial Reviews. 2005;**24**(4):271-278

[20] Kesavachandran CN, Fareed M, Pathak MK, Bihari V, Mathur N, Srivastava AK. Adverse health effects of pesticides in agrarian populations of developing countries. Reviews of Environmental Contamination and Toxicology. 2009;**200**:33-52. DOI: 10.1007/978-1-4419-0028-9_2

[21] Corriols M, Marín J, Berroteran J, Lozano LM, Lundberg I. Incidence of acute pesticide poisonings in Nicaragua: A public health concern. Occupational Environmental Medicine. 2009;**66**(3):205-210. DOI: 10.1136/oem.2008.040840

[22] Bao LJ, Maruya KA, Snyder SA, Zeng EY. China's water pollution by persistent organic pollutants. Environmental Pollution. 2012;**163**:100-108. DOI: 10.1016/j.envpol.2011.12.022

[23] Kaw HY, Kannan N. A review on polychlorinated biphenyls (PCBs) and polybrominated diphenyl ethers (PBDEs) in South Asia with a focus on Malaysia. Reviews of Environmental Contamination and Toxicology. 2017;**242**:153-181. DOI: 10.1007/398_2016_14

[24] World Health Organization: Children Face High Risks from Pesticide Poisoning. Joint note for the media WHO/FAO/UNEP. http://www.who.int/mediacentre/news/notes/2004/np19/en/

[25] Shankar S, Shanker U, Shikha. Arsenic contamination of groundwater: A review of sources, prevalence, health risks, and strategies for mitigation. The Scientific World Journal. 2014;**2014**:304524. DOI: 10.1155/2014/304524

[26] Jiang J-Q, Ashekuzzaman SM, Jiang A, Sharifuzzaman SM, Chowdhury SR. Arsenic contaminated groundwater and its treatment options in Bangladesh. International Journal of Environmental Research and Public Health. 2013;**10**(1):18-46. DOI: 10.3390/ijerph10010018

[27] Rahman MM, Sengupta MK, Ahamed S, Chowdhury UK, Lodh D, Hossain A, Das B, Roy N, Saha KC, Palit SK, Chakraborti D. Arsenic contamination of groundwater and its health impact on residents in a village in West Bengal, India. Bulletin of the World Health Organization. 2005;**83**(1):49-57. DOI: S0042-96862005000100014

[28] Rodríguez-Lado L, Sun G, Berg M, Zhang Q, Xue H, Zheng Q, Johnson CA. Groundwater arsenic contamination throughout China. Science. 2013;**341**:866-868. DOI: 10.1126/science.1237484

[29] Kim KW, Chanpiwat P, Hanh HT, Phan K, Sthiannopkao S. Arsenic geochemistry of groundwater in Southeast Asia. Frontiers of Medicine. 2011;**5**(4):420-433. DOI: 10.1007/s11684-011-0158-2

[30] Shiomi N. An assessment of the causes of lead pollution and the efficiency of bioremediation by plants and microorganisms. In: Shiomi N, editor. Advances in Bioremediation of Wastewater and Polluted Soil. Rieca: InTech; 2015. pp. 246-274. DOI: 10.5772/60802

[31] Song Y, Jin L, Wang X. Cadmium absorption and transportation pathways in plants. International Journal of Phytoremediation. 2017;**19**(2):133-141. DOI: 10.1080/15226514.2016.1207598

[32] Dhal B, Thatoi HN, Das NN, Pandey BD. Chemical and microbial remediation of hexavalent chromium from contaminated soil and mining/metallurgical solid waste: a review. Journal of Hazardous Materials. 2013;**250-251**:272-291. DOI: 10.1016/j.jhazmat.2013.01.048

[33] Li P, Feng XB, Qiu GL, Shang LH, Li ZG. Mercury pollution in Asia: A review of the contaminated sites. Journal of Hazardous Materials. 2009;**168**(2-3):591-601. DOI: 10.1016/j.jhazmat.2009.03.031

[34] Wong CS, Duzgoren-Aydin NS, Aydin A, Wong MH. Sources and trends of environmental mercury emissions in Asia. Science of the Total Environ. 2006;**368**(2-3):649-662. DOI: 10.1016/j.scitotenv.2005.11.024

[35] Lottermoser BG. Mine Wastes: Characterization, Treatment and Environmental Impacts. Berlin: Springer; 2007

[36] Schaider LA, Senn DB, Brabander DJ, McCarthy KD, Shine PJ. Characterization of zinc, lead, and cadmium in mine waste: Implications for transport, exposure, and bioavailability. Environmental Science and Technology. 2007;**41**(11):4164-4171. DOI: 10.1021/es0626943

[37] Verma C, Madan S, Hussain A, Dubey A. Heavy metal contamination of groundwater due to fly ash disposal of coal-fired thermal power plant, Parichha, Jhansi, India. Cogent Engineering. 2016;**3**(1):1179243. DOI: 10.1080/23311916.2016.1179243

[38] Waste Electrical & Electronic Equipment (WEEE) European commission, environment. http://ec.europa.eu/environment/waste/weee/index_en.htm

[39] Keller W, Heneberry JH, Dixit SS. Decreased acid deposition and the chemical recovery of Killarney, Ontario, lakes. Ambio. 2003;**32**(3):183-189. DOI: 10.1579/0044-7447-32.3.183

[40] Menz FC, Seip HM. Acid rain in Europe and the United States: An update. Environmental Science & Policy. 2004;**7**(4):253-265. DOI: 10.1016/j.envsci.2004.05.005

[41] Ojuederie OB, Babalola OO. Microbial and plant-assisted bioremediation of heavy metal polluted environments: A review. International Journal of Environmental Research and Public Health. 2017;**14**(12). pii: E1504. DOI: 10.3390/ijerph14121504

[42] Mosa KA, Saadoun I, Kumar K, Helmy M, Dhankher OP. Potential biotechnological strategies for the cleanup of heavy metals and metalloids. Frontiers in Plant Science. 2016;**15**(7):303. DOI: 10.3389/fpls.2016.00303. eCollection 2016

[43] Shiomi N. A novel bioremediation method for shallow layers of soil polluted by pesticides. In: Patil YB, Rao P, editors. Applied Bioremediation—Active and Passive Approaches. Rijeca: InTech; 2011. pp. 285-304. DOI: 10.5772/56153

[44] Hans-Jørgen Albrechtsen HJ, Smith PM, Nielsen P, Christensen TH. Significance of biomass support particles in laboratory studies on microbial degradation of organic chemicals in aquifers. Water Research. 1996;**30**(12):2977-2984. DOI: 10.1016/S0043-1354(97)84232-7

[45] Shiomi N, Inoue Y, Tomioka K, Yasuda T. Immobilization of Candida cells showing myceriu-like shapes in porous polyvinyl formal resin and their applications. Journal of Chemical Engineering of Japan. 2003;**36**(2):2977-2984

[46] Shiomi N, Yaaguchi Y, Nakai H, Fujita T, Katsuda T, Katoh S. Degradation of cyanuric acid in soil by Pseudomonas sp. NRRL-12227 using bioremediation with self-immobilization system. Journal of Bioscience and Bioengineering. 2006;**102**(3):206-209

[47] Basile A, Cogoni AE, Bassi P, Fabrizi E, Sorbo S, Giordano S, Cobianchi RC. Accumulation of Pb and Zn in Gametophytes and Sporophytes of the Moss Funaria hygrometrica (Funariales). Annals of Botany. 2001;**87**(4):537-543. DOI: 10.1006/anbo.2001.1368

[48] Ito M, Honma Y, Nakatsuka S, Komatu Y, Kawakami S, Sakakibara H. Aqueous environment conservation and metal-resource recycling technology using the moss Funaria hygrometrica. Regulation of Plant Growth & Development. 2010;**45**(1):64-72

[49] Wackett LP, Sadowsky MJ, Martinez B, Shapir N. Biodegradation of atrazine and related s-triazine compounds: From enzymes to field studies. Applied Microbiology and Biotechnology. 2002;**58**(1):39-45

[50] Govantes F, Porrúa O, García-González V, Santero E. Atrazine biodegradation in the lab and in the field: Enzymatic activities and gene regulation. Microbial Biotechnology. 2009;**2**(2):178-185. DOI: 10.1111/j.1751-7915.2008.00073.x

[51] Singh BK, Walker A. Microbial degradation of organophosphorus compounds. FEMS Microbiology Reviews. 2006;**30**(3):428-471. DOI: 10.1111/j.1574-6976.2006.00018.x

[52] Ayangbenro AS, Babalola OO. A new strategy for heavy metal polluted environments: A review of microbial biosorbents. International Journal of Environmental Research and Public Health. 2017;**14**(1). pii: E94. DOI: 10.3390/ijerph14010094

[53] DalCorso G, Fasani E, Furini A. Recent advances in the analysis of metal hyperaccumulation and hypertolerance in plants using proteomics. Frontiers in Plant Science. 2013;**4**:280. DOI: 10.3389/fpls.2013.00280

[54] Pollard AJ, Reeves RD, Baker AJM. Facultative hyperaccumulation of heavy metals and metalloids. Plant Science. 2014;**217-218**:8-17. DOI: 10.1016/j.plantsci.2013.11.011

[55] Fernando ES, Quimado MO, Doronila AI. Rinorea niccolifera (Violaceae), a new, nickel-hyperaccumulating species from Luzon Island, Philippines. PhytoKeys. 2014;**37**:1-13. DOI: 10.3897/phytokeys.37.7136

[56] Azad MAK, Amin L, Sidik NM. Genetically engineered organisms for bioremediation of pollutants in contaminated sites. Chinese Science Bulletin. 2014;**59**(8):703-714. DOI: 10.1007/s11434-013-0058-8

[57] Singh JS, Abhilash PC, Singh HB, Singh RP, Singh DP. Genetically engineered bacteria: an emerging tool for environmental remediation and future research perspectives. Gene. 2011;**480**(1-2):1-9. DOI: 10.1016/j.gene.2011.03.001

[58] de Mello-Farias PC, Chaves ALS, Lencina CL. Transgenic plants for enhanced phytoremediation—Physiological studies. In: Alvarez M, editor. Genetic Transformation. Rijeca: InTech; 2011. pp. 305-328

[59] Wang DZ, Jin YN, Ding XH, Wang WJ, Zhai SS, Bai LP, Guo ZF. Gene regulation and signal transduction in the ICE-CBF-COR signaling pathway during cold stress in plants. Biochemistry (Mosc). 2017;**82**(10):1103-1117. DOI: 10.1134/S0006297917100030

[60] Babu RP, O'Connor K, Seeram R. Current progress on bio-based polymers and their future trends. Progress in Biomaterials. 2013;**2**:8. DOI: 10.1186/2194-0517-2-8

[61] Gkorezis P, Daghio M, Franzetti A, Van Hamme JD, Sillen W, Vangronsveld J. The interaction between plants and bacteria in the remediation of petroleum hydrocarbons: An environmental perspective. Frontier Microbiology. 2016;**7**:1836. DOI: 10.3389/fmicb.2016.01836. eCollection

[62] Rajkumar M, Vara Prasad MN, Freitas H, Ae N. Biotechnological applications of serpentine soil bacteria for phytoremediation of trace metals. Critical Reviews in Biotechnology. 2009;**29**(2):120-130. DOI: 10.1080/07388550902913772

[63] Xie X, Ye M, Hsu P-C, Liu N, Criddle CS, Cuib Y. Microbial battery for efficient energy recovery. Proceedings of the National Academy of Sciences of the United States of America. 2013;**110**(40):15925-15930. DOI: 10.1073/pnas.1307327110

[64] Timmer J. "Microbial battery" could turn wastewater into electricity: A hybrid of fuel cell and battery with bacteria supplying the electrons. ars technica. 2013. https://arstechnica.com/science/2013/09/microbial-battery-could-turn-wastewater-into-electricity/

[65] Newton GJ, Mori S, Nakamura R, Hashimoto K, Watanabe K. Analyses of current-generating mechanisms of Shewanella loihica PV-4 and Shewanella oneidensis MR-1 in microbial fuel cells. Applied Environmental Microbiology. 2009;**75**(24):7674-7681. DOI: 10.1128/AEM.01142-09

[66] Lee S, Kim DH, Kim KW. The enhancement and inhibition of mercury reduction by natural organic matter in the presence of Shewanella oneidensis MR-1. Chemosphere. 2017;**194**:515-522. DOI: 10.1016/j.chemosphere.2017.12.007

Chapter 2

The Pollution of Water by Trace Elements Research Trends

Khaled Al-Akeel

Additional information is available at the end of the chapter

http://dx.doi.org/10.5772/intechopen.72776

Abstract

Water pollution has been a growing issue for the last decades. This is mainly because of the boost in human population, and the motivations that lead to technological advances for the welfare of the society. Water pollution originates from different sources such as agricultural, municipal, industrial, and landfills drainage waters. These pollutants, which are either organic, nutrient, or heavy metals pollutants, are very deleterious to the natural ecosystems and eventually harmful to humans. Different procedures have been proposed for handling heavy metals water pollution, which encompass electro-osmosis, ion exchange, electrokinetic, sludge activation, as well as phytoextraction. Water contaminants are also removed using flotation, membrane filtration, aeration, precipitation, coagulation—flocculation, ion exchange, and electrochemical treatment. These procedures are costly and have prompted the use of other techniques, such as phytoremediation. Phytoremediation involves the utilization of plant species to alleviate the impacts of environmental pollution. It could be implemented to eliminate pollutants from various natural ecosystems including water, soil, and air or to develop new vegetation growth on disturbed or barren ground. Different plant species have been used for phytoremediation. This chapter addresses trace elements pollution of natural water resources in details and the abilities of Aquatic plant communities such as Reed plants (*Phragmites australis*) to absorb soluble trace elements from water.

Keywords: water pollution, trace elements, research trends, phytoremediation, *Phragmites australis*

1. Introduction

Water pollution is one of the biggest challenges that humans face in recent years. The control of water and soil pollution has been tackled in many studies conducted worldwide to find ways and means to eradicate this major problem. Different methods for extracting pollutants were developed to be used at the commercial level. With the advancement of technology, extraction techniques are more rapid than before, however radioactive, biological as well as toxic pollutants greatly increased to levels, which render natural water resources toxic especially if the pollutants accumulate beyond certain levels. This chapter addresses trace elements pollution of natural water resources in details and the abilities of Aquatic plant communities such as Reed plants (*Phragmites australis*) to absorb soluble trace elements from water.

The Reed plants are commonly growing in different habitats throughout the world ranging from deserts, coasts, to mountains. The Reed plant poses a great problem when they invade fertile farm lands and causes water leaching. These plants possess potential efficiency for heavy metals uptake via their roots and shoots during their growth. The use of phytoremediation (with Reed plants) as a cheap and clean technology, encouraged numbers of researchers to develop new methods to remove trace elements from the natural ecosystems. Researchers have constructed large wetlands using reed plants to remove pollutants [1, 2]. These research experiments included wastewater, sludge, and sewage as well as industrial waste water, where aquatic plant species beds were used for water purification [3].

Trace toxic elements include Cu, Cd, Pb, Cr, As, and Hg, which are needed for living things within certain levels, and if more concentrations of these metals are taken, they are harmful to living organisms. These elements are generally not toxic in their metallic form, but when they are present in fine powder form, they can toxic if inhaled or ingested. Usually, their compounds with other elements are poisonous to living organisms [4].

Copper, Chromium, Cadmium, and Lead are among the main pollutants of water resources that have densities more than 6 g/cm^3. Cu, Cd, Pb, Cr levels in the soil solutions fluctuate between 1000 ppm and few. However, manganese level in the soil solution ranges between 20 and 10,000 ppm. Some kinds of soils may have higher concentrations of these trace elements. When the concentrations of these trace elements extremely higher levels in the soil environment, the soil is said to be polluted with heavy metals and cannot be used for agricultural crop production. The increase in the concentration of heavy metals in the soil environment could be attributed to human activities, such as the production of energy, agricultural production, and mining [5–7]. All these activities cause the buildup of heavy metals concentrations in the soil solution which eventually, ends up in the tissues of the different plant and animal species as well as human [8], and heavy metals may accumulate to toxic levels in the ecosystem [9].

The application of both chemical and animal fertilizers to crop plants increased the concentration of heavy metals greatly in the soil solution, thus polluting underground water resources. In addition, the accumulation of plant nutrients, such as nitrogen (N) and phosphorus (P), due to the high rates of fertilizer application to crop plants, in surface water resources lead to what is known as Eutrophication. Eutrophication results in an increased growth rate of plant species, e.g., algae. In addition, the extensive use of fertilizers in crop production leads to heavy metals buildup in the different water resources [10]. The high accumulation rate of chemical substances and heavy metals causes a notable threat to both human and animal health [10, 11].

2. Water pollution

Water pollution scale: Water pollution is mainly attributed and distributed by human activities in the agricultural sector as compared to the industrial sector [12]. Research and development in the field of water pollution is focused on the remedial techniques worldwide in the last three decades. Water pollution can happen due to accidental, carelessness, or illegal discharge of polluted wastewater in the natural surface water resources. Such incidents occur more due to the lack official regulations, monitoring.

Procedures and public awareness about the consequences of water pollution [13, 14].

3. Water pollution sources and implications

The main sources of the pollution of water resources are the drainage of wastewater from cropped farms, factories, and municipalities that is defined as **point source pollution**. The amount of this type of water pollution decreases as the distance from the point source is increased and it can easily be monitored and controlled. However, in **diffuse pollution source**, pollutants get in water from diversified sources such as atmosphere, agricultural chemicals, i.e., fertilizers and pesticides, drainage, sewage industrial wastewater, infiltration of ground water. The impacts of pollution from diffuse source could be troublesome. Most methods are developed to treat point source pollution. Water pollution exhausts the concentration oxygen dissolved in water sources and influences both pH and temperature levels that must be stable for aquatic organisms survival in the ecosystem [12]. Oxygen is considered as a limiting factor for the growth and development of fish, which need high amounts of oxygen for their metabolic activities; however, the increase in temperatures levels will reduce the oxygen solubility in water and thus reducing its concentration to meet fish needs [15, 16].

3.1. Organic pollutants

This is considered one of the ancient water pollution kinds when organic matter is discharged into water sources. The main source of organic pollutants is farm's drainage water that contains animal waste and agricultural chemicals (pesticides, herbicides, and fertilizers), domestic wastewater, and industrial water. Suspended organic matter constitutes the main percentage of organic wastes and it can be monitored as the levels of oxygen concentration [14, 17]. The main components of an organic waste typical composition are lignin, cellulose, lipids, protein, amino acids, and ash [18]. The chemical reactions that happen to these materials, while they are present in water resources could lead to the depletion of water oxygen content that may finally have a negative impact on the ecosystem. The strength of wastewater that causes the organic pollution is usually measured by its biological oxygen demand (BOD), which is expressed as mg/liter.

3.2. Nutrient pollutants

There are several factors that are detrimental to plant growth, these factors include the availability of macro and micro nutrients, e.g., nitrates, phosphates, etc. [12, 19], when the discharged organic pollutants are decomposed, they make available phosphates, nitrates and other nutrients for plant growth. However, the high growth rate of plant species may have negative effects on the aquatic environment. The negative impacts include [12]:

1. The physical alterations in the habitat may cause faunal changes in the populations of filamentous and macrophytes algae.
2. The dissolved oxygen concentrations may be depressed due to the respiration of the densely growing aquatic plant species.
3. Algal species bloom may cause several problems regarding water qualities, and the water might tend to be poisonous to aquatic animal species, including fish.
4. Plant species biomass decay may create huge amounts of organic sediments that may adversely affect the aquatic ecosystem.

3.3. Heavy metal pollutants

The chemical pollutants may include both organic and inorganic materials, although many inorganic substances, such as metals, are considered important for the well-being and survival of living organisms in any environment. High or low metal concentrations can have negative impacts on the environment. There are two major groups of toxic metals: (a) Mercury (Hg), Lead (Pb), Cadmium (Cd), and Arsenic (As) which are non-essential and (b) Zinc (Zn), Chromium (Cr), Copper (Cu), and Nickel (Ni) which are essential to plant and animal species growth and development. These trace elements may give rise to soil, water, and air pollution, which has been studied by many scientists [13].

Pollution resulting from human activities produces wastes rich in metals content which rises their concentrations in the natural ecosystem, especially water resources to toxic levels. The major human activities that lead to water pollution include industrial wastewater, agricultural drainage water, sewage water, and oil pollution. Water pollution is considered a major problem that causes many diseases all over the world. About 14,000 individuals lose their life daily due to heavy metal water pollution [20, 21].

The developments in the technological and industrial sectors depend on the use of the minerals natural source and the aquatic ecosystems are considered as the final recipients of their toxic, solid as well as drainage waste water. The national water quality report of United States of America, for example, indicates that 45% of tested stream miles, 47% of tested lake acres, and 32% of bay and estuarine square miles are described as polluted [22].

The protection of human health and aquatic ecosystem well-being is a fundamental concern all over the world, and it has been deem necessary that ways for the management of aquatic resources should be considered within the dynamics of the ecosystem conditions and their exploitation for human needs continues to be sustainable [23].

The exposure of individuals to toxic elements, such as Arsenic, Cadmium, Chromium, Mercury, and lead, can result in serious health hazards [24]. The sources of these toxic elements include mainly chemical industries drainage wastes, medical waste, excessive use of fertilizer and their production, oil refineries, and pharmaceutical industries. These waste waters carry sulfide, pyrite, pyrhotite converting into sulfate into ground water, rivers, and lakes [25] (**Table 1**).

Element	Max. admissible conc. (mg/dm³)	Max. permissible conc. (mg/dm³)
Cd	0.005	0.01
Cr	0.05	0.10
Cu	0.05	15.00
Pb	0.05	0.10
Zn	5.0	15.00
Source: [33].		

Table 1. The WHO admissible and permissible levels of trace elements in fresh water.

Many different studies in the fields of heavy metals pollution [26–32] have covered topics such as:

1. Constrains of soil and water environmental standards.
2. Environmental conditions and pollutant sources.
3. Pollutant concentration levels.
4. Chemistry of toxic water pollutants.
5. Solubility, adsorption, dispersion, volatility, and movement of chemical pollutants.
6. Hydrogeology of polluted sites.
7. Monitoring of the quality of groundwater.
8. Chemistry and mechanics of soil pollutants.

The exposure for a long period of time to different concentrations of heavy metals may lead to grave health and environmental consequences and some of these metals are deleterious, such as As, Pb, and Hg. The trouble regarding these metals is the fact that they do not degrade or accumulate, but can be transmitted from one environmental site to another, these facts for their inclusion, handling, sweep, and drainage Ho and El-Khaiary, Long et al., Nriagu [34–36].

4. Water pollution treatment method

The researchers, government authorities, and businessmen have endeavored to tackle heavy metals pollution of water, air, soil, and its impacts on human bodies. Many technologies and regulations have been formulated to deal with the monitoring and determination of heavy metals pollution and convenient arrangement are taken [37, 38].

Different laboratory and field mechanisms have been proposed for handling heavy metals pollution [39], which included electro-osmosis [40], ion exchange [41], electro-kinetic [42], sludge activation [43] as well as phytoextraction [44–47]. Water contaminants are classically removed using different procedures such as flotation, membrane filtration, aeration, precipitation, coagulation-flocculation, ion exchange, and electrochemical treatment [34, 48]. These classical procedures are thoroughly discussed in the literature [49, 50]. However, these procedures are costly and have prompted the use of new techniques, such as phytoremediation. The different new techniques that are used for the control of heavy metal pollution include immobilization, soil flushing, electro-kinetics, phytoremediation, etc. [51].

4.1. Phytoremediation

Phytoremediation is the process in which plants are used to get rid of environmental pollutants. It takes into account all physical, chemical, or biological approaches that use plant species for decontamination of polluted sites. Two basic principles applied in this technique are phytoextraction (harvesting) and phytostabilization (root fixing) [13, 52–55]. Aquatic plant species are capable for the extraction and accumulation of heavy metallic pollutants without severely affecting their growth in the polluted site [56, 57], that nominates them for the treatment of heavy metals pollution [58]. Using phytoremediation to extract heavy metals from both natural water as well as soil resources is a feasible economic and non-ambiguous action [59]. The aquatic plant species *Phragmites australis* is a perennial grass that naturally grows in wet habitats throughout the world. This plant can tolerate environmental stresses such as heavy metal pollution [60]. *Phragmites australis* plants are widely used in the construction of artificial wet sites for the treatment of industrial drainage water rich in its heavy metal contents [56].

4.2. Rhizofiltration

Rhizofiltration is one of the phytoremediation mechanisms that employs the roots of plant species in the uptake, accumulation, and sedimentation of heavy metals from drainage water and contaminated soils. Rhizofiltration utilizes overland plant species because these plants are characterized by longer and fibrous root systems that are covered with root hairs, which increase their absorptive, surface areas [46, 61]. In this phenomenon, metal uptake does not encompass any biological procedures [13, 62]. *Eichhornia crassipes* plants can accumulate Cd and Pb at a concentration of more than 6000 ppm in plant tissues [63, 64]. *Azolla filiculoides* Lam water Fern can uptake Cd, Cu, Ni, and Zn from the soil solution at levels of 10,000, 9000, 9000, and 6500 ppm, respectively [65, 66]. *Nelumbo nucifera* Gaertn and *Nymphaea alba* L. plants can accumulate Cr up to 3000 mg Cr kg [67]. Severe toxic effects on health are observed at low concentrations of 0.5 mg/L for Cr (VI). Cr (VI) is considered as lethal for human being at a dose higher than 3 g same as an example of Cr, other metals Cu, Cd, and Pb also cause high degree of pollution based on their toxicity and their ability to generate different compounds in water. Thus, it can be noted from this discussion that heavy metal pollution of soil and water can be treated using plant species. The drastic permanent solution can be obtained if we choose the right plant species to be used after knowing the actual causes of water and soil pollution.

Government personnel, scientists, and the public are concerned about the ever growing ecological threats that include global warming, the diminishing natural resources that promoted intensive research for the development of new techniques for removing heavy metals from the polluted sites, such as using plant species, improving existing and new decontamination procedures

[13, 68, 69]. The problem is also amplified in advanced as well as developing countries, as a result of the ever increasing industrialization and auricular development releasing high amount of toxic materials if their drainage water into the existing natural ecosystems [56, 68, 70, 71].

All these consequences have promoted the use of phytoremediation for the removal of toxic pollutants, because aquatic plants grow and flourish naturally in drainage canals [2]. Phytoremediation, using aquatic plants such as Reed plants, provides a chance for its use as a feasible non-destructive technology to eliminate pollutants from soils or water ecosystems. This was assured by the work of many scientists who analyzed aquatic plant tissues for their mineral contents [55, 72, 73].

5. Reed plants characteristics

Reed (*Phragmites australis* (PA)) plant is a perennial aquatic plant species that has given wide attentiveness for treating heavy metal polluted soils and water sites due to the fact that it can tolerate the adverse effects of high levels of toxic metals, without affecting its growth and high yield rates, PA can naturally grow well in natural or artificial sites polluted with Zn, Ni, Pb, As, and Cd [3, 74]. It is recorded growing naturally in low-level coastal plains or river-flooded areas in North America, Europe, Middle East, Africa, and Australia.

The reed plants can grow to a stem length of more than 350 cm and is characterized by its efficiency to move O_2 from its shoot zone to its root system. The root system can develop and extend even in water-logged environments [65, 66]. Kilkuth in Germany used Reed beds for drainage water treatment in 1970s and in England in 1985 [75].

6. Heavy metal pollutant control methods

Lower concentrations of Copper, Selenium, and Zinc are essential for human body's metabolism, but at higher concentration, these metals (Fergusson [17]) as well as Mercury, Cadmium, and Lead are toxic and can eventually cause a number of diseases that lead to muscular, neurological, and physical degenerative impacts. In recent years, different procedures were developed to eliminate metal pollutants from drainage and drinking water [12]. These procedures are discussed as follows.

6.1. Chemical precipitation

It is considered as one of the widespread procedures for the removal of metal ions from water solutions [76]. The methodology involves the production of metallic hydroxides, e.g.

$$M^{n+} + n\,(OH-) \rightarrow M\,(OH)_n. \qquad (1)$$

where M (OH) is the metal insoluble hydroxide, Mn^+ is the dissolved metal ions, and OH^- is the precipitant.

The main hindrances of this procedure are the high amounts of chemicals used, and excessive wastes that are produced which need to be eventually treated. In addition, cumbersome metal deposition, fiddling sediment of hydroxides, assemblage of metal depositions, and long-term effects of the disposal of wastewater also need to be taken care of [77, 78].

6.2. Coagulation-flocculation

Coagulation and flocculation happen in successive operations in order to upset the stability of the unsettled particles, to permit particle impact and to regulate the growth of aggregates. Both operations must be achieved to accomplish the elimination of pollutants [79]. Coagulation, fore mostly, takes place to upset the stability of the particles leading to their precipitation. Flocculation increases the size of the particles through the aggregation of unsettled particles. This is mainly accomplished by pH modification and addition of ferric or aluminum compounds to control the dissonance within the colloidal particles. The use of lime, as coagulating agent, improves the settlement of the sludge, reduces watering, and makes it capable for controlling bacterial activities. The main drawback of the use of lime is the high costs [78].

6.3. Flotation

This methodology is effective for liquids as it uses bleb connection to split up solids. There are five kinds of flotation, i.e., biological-, electro-, vacuum-air-, dissolved-air-, and dispersed-air flotation. The most widely used one for metal removal from drainage water is dissolved-air flotation [79].

6.4. Aeration

Aeration is needed when the water has oxygen shortage condition due to municipal, agricultural, and industrial drainage water discharge. Aeration is conducted by introducing air at the bottom of the water reservoir or by surface stirring generating a basal device that allow air and water mixing which would help in releasing and removing of harmful gasses, e.g., hydrogen sulfide, carbon dioxide, and methane [78]. It can be utilized to handle both drainage and drinking waters. It can be classified into surface, sub-surface, and natural aeration [79].

6.5. Membrane filtration

This technique is beneficial because no chemicals were used, and has a comparatively minimal energy use and it can easy be conducted in gradual stages. This process can be used to remove dissolved pollutants, e.g., organic compounds, suspended solids, and heavy metals. There are different sorts of filtration mechanisms that can be used, based on the targeted particle size that is required to be removed. These mechanisms include reverse osmosis (RO), ultrafiltration (UF), and nanofiltration (NF). Ultrafiltration uses permeable membranes with pores ranging between 5 and 20 nm depending on the material to be extracted. Nanofiltration technique depends on the steric effect as well as the electrical effect. It utilizes the electrical charge differences between the membranes' anions and the pollutants' cations that results in repulsive forces to separate these metallic pollutants. In reverse osmosis, the exerted pressure forces keep the trace elements and clean water can be collected from the second direction of the membrane. Reverse osmosis is more efficient as compared to ultrafiltration and nanofiltration mechanisms, since it extracts 97% of the targeted heavy metal in the 20–200 mg/L range [13, 78].

6.6. Ion exchange

The exchange of ions of the same charge between an insoluble solid and a solution in contact with it used in water-softening and other purification and separation processes. Thus, metals can be extracted from solution with the help of suitable reagents [13]. This procedure is effective when pH of the solution (polluted water source) ranges from 2 to 6 and demands the removal of the suspended solids by other methods prior to the implementation of the ion exchange method [79].

6.7. Electrochemical treatment

This procedure performs metal extraction using the combination of membrane and ion exchange mechanisms. The main methodology is accomplished by passing polluted water through an ion exchange membrane. This membrane consists of thin plastic materials that have either cationic or anionic electric charge. This methodology is efficient, although it is expensive as far as chemicals, sludge handling costs, and its high-energy use [13].

6.8. Microbial biosorption

The utilization of non-living microbial tissues for retaining heavy metals, which is not a metabolic activity, is defined as biosorption [80]. The microbial cells are good biosorbents because they have a high surface area, and thus have high number of biosorption sites. Industrial activities, such as electroplating and mining, produce high amounts of biomass that could be used in the removal of heavy metal pollutants [25]. Biomass can be also been grown using fermentation techniques and inexpensive growth media. Dead cell biomass can be more beneficial as compared to live cell biomass, as systems utilizing living cells can be affected by metal ion concentrations, temperature, pH, and constant requirement of nutrient supply for living cells [13, 81]. The industrial waste when fermented can be processed into biomass. Microbial cells in the biomass can be removed by many methods such as heat treatment, autoclaving, and vacuum drying, using acids, alkalis, detergents, organic compounds, and mechanical disruption.

The effective amount of absorbing material to be used can be estimated by comparing its strength to remove the sorbent material from polluted water with other substances reported in other studies. The amount is defined in terms of the amount of heavy metal extracted (milligrams) per amount of sorbent used (gram). The biosorption methodology is affected by pH, temperature as well as the ion concentration in the solution [80].

6.9. Phytoremediation

Phytoremediation is a technique that uses plant species and associated-soil micro-organisms to extract metals present in the ecosystem, i.e., air, soil, and water [82, 83]. Phytoremediation is now utilized for treating pollutants, e.g., heavy metals, fertilizers, pesticides, chlorinated solvents, petroleum hydrocarbons, explosives, etc. All other traditional methods for the removal of heavy metal pollutants from the natural ecosystems are costly and use large quantities of chemicals and create waste treatment problems, even if these procedures are advantageous for their rapid extraction of pollutants in small sites [17].

Phytoremediation surpassed the traditional procedures because it helped in maintenance of biological activity, site restoration, and partial decontamination as well as it is unobtrusive, has a possibility of bio-recovery of metals and not costly [47, 84]. Due to these merits, phytoremediation is enumerated as a "green," sustainable pollution remedy procedure. Phytoremediation has been classified into five divisions:

i. Phytoextraction: plants uptake metals and accumulate them in the usable tissues [85];

ii. Phytodegradation: plant species and associated micro-organisms break down organic pollutants [86, 87];

iii. Rhizofiltration: the plant's root system absorbs metals from polluted water sources [46, 51];

iv. Phytostabilisation: plant species reduce pollutants bioavailability in the natural ecosystems by fixation or by prohibition of their migration [88] and,

v. Phytovolatilization: volatilization of pollutants into the atmosphere through plants [86, 89]. The next section explains a detailed review of this emerging green technology.

vi. Phytovolatilization: the evaporation of pollutants into the air via plants [86, 89].

The advances in the fields of research and technology have promoted the growth of different industries resulting in matchless unrests in the ecological cycles [78]. The recent admission of toxic chemicals and resettlement of natural substances into different ecosystems (soils, water, and air) have given rise to major demands of self-purifying capacity of the natural ecosystems [90, 91].

The present off-site procedures for treating polluted water resources include extracting and treating of pollutants using adsorption capacity of activated carbon, micro-organism or air denudation; while on-site procedures for treating polluted water resources include stimulation of aquifers aerobic and anaerobic micro-organisms activities. All predominant procedures are expensive and require high technical human resources to execute. Thus, stakeholders are looking forward for the development of cheaper and speedy techniques for treating highly contaminated water resources, wetlands, and soils [13].

It is claimed that about 80% of the contaminated soil water is found within the top 20 m of the soil depth. This indicates that the removal of water pollutants can be performed using the cheap phytoremediation technology [92]. Many experimental studies have been conducted on the use of plant technologies in treating contaminated natural resources [93–96]. The various phytoremediation methods encompass: the adjustment of the physical and chemical characteristics, liberation of the root exudates, amend aeration and allow more of the oxygen to reach the root zone, oppose and reduce the transport of chemicals, affect the shared metabolic activities of plant and micro-organisms' enzymes, and deduce the vertical and horizontal movement of contaminants [97]. When the water pollutants, i.e., heavy metals, are found in low concentrations, phytoremediation may prove to be the most feasible technique for their treatment due to its cheaper costs and its efficiency [98].

The main phytoremediation procedures can be classified as on-site, in a living organism and off-site types. On-site phytoremediation is the modest and cheaper of all three classes and the plants are growing in direct contact with pollutants.

http://dx.doi.org/10.5772/intechopen.72776

There are numerous technicalities by which plant species can rectify water and soil pollution. During phytoremediation, the merits of the technique can be influenced by physical and chemical pollutants traits (molecular weight, vapor pressure as well as their solubility in water), environmental traits (organic matter in the water solution, pH as well as temperature), and plant traits (root system characteristics and enzymes) [90].

7. Use of plants for the treatment of pollutants

Many plant species have capabilities to uptake and accumulate high levels of metallic organic compounds without any toxic impacts, such as reed plants (*Phragmites australis*), Indian mustard (*Brassica juncea* L.), willow (*Salix* species), poplar tree (*Populus deltoides*), Indian grass (*Sorghastrum nutans*), sunflower (*Helianthus annuus* L.), water hyacinth (*Eichhornia crassipes*), channel grass (*Vallisneria spiralis*), alfalfa (*Medicago sativa*), brassica (*Brassica napus*), kenaf (*Hibiscus cannabinus* L.), tall fescue (*Festuca arundinacea* Schreb), bermudagrass (*Cynodon dactylon*), and barley (*Hordeum vulgare* L.) [13, 99].

These plant species have attained heavy metal tolerance characteristics that modified them to survive in highly heavy metal polluted ecosystems [13, 99], and they show high strength to accumulate heavily metallic ions, such as nickel, zinc, copper, chromium, and even radionuclides. The threshold for hyper-accumulation of heavy metals is known as the accumulation in plant tissues of concentrations greater than 0.1% on dry weight basis and not less than 0.01% for cadmium. Many types of plant species could be utilized in phytoremediation [61]. These plant species have specific characteristics that assist in their normal growth in polluted sites and at the same time uptake and accumulate heavy metals in their roots and shoots [100, 101]. Reed (PA) plants are plant species that are widely dispersed and are found in wide range of habitats worldwide [102]. PA plants are aquatic perennial grass [103]. A number of studies have been conducted on PA plants growth and development [104], mineral content [105], response to heavy metals [49, 104, 106–108], response to salt stress, [108, 109] as well as the different reed plant subspecies phenotypic, genotypic, evasive characteristics [49, 96, 109]. The results on the available research studies indicated that reed plant species significantly differ in their response to salt stress [102, 109, 110], this could be due to genotypic differences [103, 111]. This is expedited by limited intake and conveyance to the shoots of sodium, sulfur, and to some extent Chlorine, the capability of maintaining a comparatively high photosynthetic rate, the capability of promoting water use efficiency, and proline production of osmotic adjustment. NaCl and Na_2SO_4 influence the water and gas operations during photosynthesis. NaCl is more poisonous to plants than Na_2SO_4 [109].

The reed plants can tolerate high-environmental stresses due to their ecological, physiological as well as morphological characteristics and will develop normally in heavy metal (Zn, Cd, and Pb) polluted ecosystems [112]. These plants possess high phytoremediation as well as detoxification traits and have been excessively utilized in the construction of artificial wetlands for the treatment of heavy metal pollution of industrial drainage water [113]. With the recent improvements in the application of scientific knowledge of phytoremediation for practical purposes, especially in pollution eradication and the search for greener alternatives for pollution enucleation, the interest has been directed toward reed plants reactions to heavy metal environmental stresses [114, 115]. However, how reed plant cells become tolerant to toxic heavy metal levels and why they are hyper-accumulator of these metals is unknown [114].

Tylova et al. [104] compared *Phragmites australis* and *Glyceria maxima* plants, which are plants adapted to growing in saline conditions, as in a salt marsh. They found that the nitrogen accumulation of the two plant species in their underground tissues did not worth considering. The annual production of the two plant species was the same, but they differed in the allocation of biomass between their rhizomes and roots. Reed plants assigned more dry weight to rhizomes as compared to Glyceria. This is demonstrated in **Figure 1**.

Reed plants were found to be more tolerant to N, P, K, ammonium compounds as well as to salt stresses as compared to Glyceria plant [108]. This was reflected by the growth of reed roots to greater soil depth, where the accumulation of NH_4^+ ions was evident, which improved reed plant tolerance to mechanical damage, mowing, and grazing [108]. Unamuno et al. [107] investigated reed plants' copper adsorption and reported that initial adsorption of Cu was high in the plants vegetative litter; but after 50 min, it decreased drastically. If the reed plants shoots were not harvested from time to time, then the plants could accumulate high amounts of the toxic metals and when they get dry, the vegetative parts would decompose and raise the waste water toxicity [116, 117]. During the movement of the cations from the soil solution to the root systems of the plants, only a small percent of free hydrated ions occur and their movement into the cell require metabolic reactions between the two media [118, 119]. The metal intake capability can be improved by root activities as they liberate solubilizing enzymes that stimulate this method in the rhizosphere [120]. Roots also liberate organic acids, flavones, nucleotides, amino acids as well as sugars [13]. This method is also affected by constancy of different metals chelates under different pH values because factors released by root cells are normally pH specific in the chelating process [121]. The intake of cations by the root system cells depends also on their concentration in the soil solution [12, 17, 122].

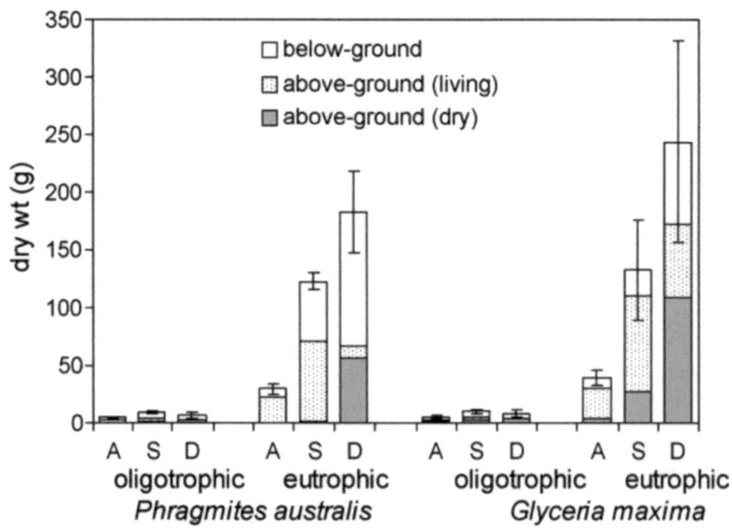

Figure 1. Below ground tissues development in reed and Glyceria plants source: Tylova et al. [104].

Vymazal et al. [49] claimed that heavy metals levels were reduced in the order of roots > rhizomes ≥ leaves > stems of the reed plants and that the heavy metal (Cd, Cr, Cu, Ni, Pb, and Zn) levels in the shoots and roots plant tissues were the same as those reported for plants growing in natural ecosystems. However, these levels were much less than for plants that were irrigated with mines, smelters, or highway drainage waters. The mean leaf to stem ratio and root to leaf ratio of metal content was found to reach 1.5. The metal uptake level can greatly be increased by increasing the nutrient availability to reed plants. These plants usually have high growth rates during the growing season even when planted from the seeds or rhizomes. This indicates that the handling of reed plant seedlings can be more discreet without paying any attention to the present situation of eutrophication [106].

Bragato et al. [123] determined the uptake of Zn, Cu, Cr, and Ni in a constructed wetland and reported that the distance from the inlet had no impact on the shoot dry weight or N, P, K, Na, Zn, Cu, Cr, and Ni contents in the aerial parts of the plants. With the exception of Na, reed plants adsorbed more of these elements than *Bolboschoenus maritimus* Reed plants heavy metal accumulation was found to increase at the end of the season. Heavy metal level in the arriving water as well as in the soil was not correlated to the amount of heavy metals of the growing plants.

8. Heavy metal accumulation in plant species

A metal assemblage depends on the ability of plant cells to uptake minerals and the number of binding sites present in the intracellular spaces. Drainage water or water contaminated with micro- and macro-organism has complications regarding specific cells and tissues differences, within cells transport level, consanguinity of chelating particles, as well as transport efficacy that can influence the rate of mineral accumulation [118], as well as operations like movement and uptake of ions from the soil solution, assortment and retention inside the root system, effectiveness of xylem for loading and transmitting of ions, allocation of ions in the different ion sinks in the shoot, confinement and pilling of the various ions in leaf cell [49].

By piling up of heavy metals, strange substances in ecosystem and radioactive nuclides via phytoremediation traits of plant species, the treatment of polluted water and soil resources will be feasible. When heavy metals have been accumulated in plant tissues, they will be turn into save metabolites. In addition, the heavy metals can be transported to plant shoots that could be harvested. The harvested material could be dried and then ashed or composted. During these processes, the amount of waste produced would be minimal when compared with other mechanisms of waste disposal [124]. Plant species that accumulate high concentrations of various cations are of high biogeochemical and environmental research importance.

9. The commercial use of reed plants for phytoremediation

Phytoremediation is a technique for treating polluted sites and it does not cause any damage and it is cheap, however it has not been utilized fully at commercial basis. Many USA

Site	Plant species	Pollutants	Applied Procedures
Trenton, NJ	*Brassica juncea*	Lead	Phytoextraction
Anderson, SC	*Populus deltoids* × *P. balsemifera* (hybrid polar) grasses	Several heavy metals	Phytostabilization
Beaverton, OR Landfield reclamation [20]	*Populus* spp. (cottonwood)	Unspecified	Vegetative cover/water pollution prevention
Ketowice, Poland	*Brassica jimcoa*	Cadmium and lead	Phytoextraction
Switzerland (landfield)	*Salix viminalis*	Cadmium, copper, and zinc	Phytoextraction
United Kingdom [43]	*Salix* spp.	Nickel, cadmium, copper, and zinc	Phytoextraction Phytostabilization
Hlemyzdi, Czech Republic [43]	*H. annuus*, *C. sativa*, *Z. mays*, and *C. hallery*	Zinc	
Dumach, Switzerland [43]	Improved *Nocotiana* spp. Plants (tobacco)	Cadmium, copper, and zinc	
Lommel, Belgium [43]	Grasses	Cadmium, lead, copper, and zinc	Phytostabilization
Balen, Belgium [43]	*Brassica napus*	Cadmium, lead, and zinc	Phytoextraction

Source: Gardea-Torresdey et al. [124], p. 1807.

Table 2. Commercial application of phytoremediation in USA and Europe.

and European pilot studies indicated that phytoremediation could be performed on large-scale basis. Most of the plant species that are utilized in phytoremediation are trees or herbaceous plants (**Table 2**). Most of these plants have been used in restoring heavy metal polluted sites. The USA market for phytoremediation may exceed 100 million dollars [51, 125]. Phytoremediation, as a technique, is advancing; but according to Glass [125], it would be accepted through social processes and information dissemination about its potential use in removal of specific contaminants. Gardea-Torresdey et al. [124] claimed that phytoremediation could constitute the main technique for handling of toxic wastes in polluted sites in the near future.

10. Conclusion

This chapter provided review of theoretical and practical views regarding heavy metal pollution of soil and water resources, methods to treat heavy metal pollution, phytoremediation concepts, and uses of various plant species in phytoremediation. However, it is not known that metals can be easily be extracted by the different plants and the capabilities of various plants as hyper-accumulator to which heavy metals.

Author details

Khaled Al-Akeel

Address all correspondence to: kaalakeel@kacst.edu.sa

Life Sciences and Environment Research Institute (LSERI), King Abdulaziz City for Science and Technology (KACST), Riyadh, Saudi Arabia

References

[1] Lee B, Scholz M. What is the role of Phragmites Australis in experimental constructed wetland filters treating urban runoff? Ecological Engineering. 2007;**29**:87-95

[2] Lu Q, Zhenli LH, Graetz DA, Stoffella PJ, Yang X. Phytoremediation to remove nutrients and improve eutrophic stormwaters using water lettuce (*Pistia stratiotes* L.). Environmental Science and Pollution Research. 2010;**17**:84-96

[3] Miao Y, Xi-yuan X, Xu-feng M, Zhao-hui G, Feng-yong W. Effect of amendments on growth and metal uptake of giant reed (*Arundo donax* L.) grown on soil contaminated by arsenic, cadmium and lead. Transactions on Nonferrous Metals Society of China. 2012;**22**: 1462-1469

[4] Duffus J. Heavy metals - a meaningless term. Pure and Applied Chemistry. 2002;**74**(5): 793-807

[5] Blaylock MJ, Huang JW. In: Raskin I, Ensley BD, editors. Phytoremediation of ToxicMetals: Using Plants to Clean up the Environment. New York: John Wiley and Sons Inc.; 2000. p. 53

[6] Alloway BJ. Heavy Metals in Soils. Glasgow: Blackie Academic & Professional; 1995

[7] Woolhouse HW. In: Lange OL, Novel PS, Osmond CB, Ziegler H, editors. Encyclopedia of Plant Physiology. vol. 12C. New York: Springer Verlag; 1981. p. 246

[8] Sager M. In: Stoeppler M, editor. Hazardous Metals in the Environment. Amsterdam, The Netherlands: Elsevier Science Publisher; 1992. p. 133

[9] Ensley BD. In: Raskin I, Ensley BD, editors. Phytoremediation of Toxic Metals: Using Plants to Clean up the Environment. New York: John Wiley and Sons Inc.; 2000. p. 3

[10] Ongley ED. Control of Water Pollution from Agriculture, Food and Agriculture Organisation of United Nations (FAO – UN). Irrigation and Drainage paper no. 55; 1996

[11] Wang YJ, Lin JK. Estimation of selected phenols in drinking water with in situ acetylation and study on the DNA damaging properties of polychlorinated phenols. Archives of Environmental Contamination and Toxicology. 1995;**28**:537-542

[12] Abel PD. Water Pollution Biology, 2/e. London: Taylor and Francis; 2002

[13] Wang LK, Chen JP, Hung Y, Shammas NK, editors. Heavy Metals in the Environment. Boca Raton: CRC Press; 2009

[14] Mackenthun KM. The Practice of Water Pollution Biology. Honolulu: University Press of the Pacific; 2005

[15] Varley M. British Freshwater Fishes: Factors Affecting their Distribution. London: Fishing News; 1967

[16] Alexander R. Functional Design in Fishes. London: Hutchinson; 1970

[17] Fergusson JE. The Heavy Elements: Chemistry, Environmental Impact and Health Effects. Oxford: Pergamon Press; 1990

[18] Higgins IJ, Burns RG. The Chemistry and Micro-Biology of Pollution. London: Academic Press; 1975

[19] Moss B. Ecology of Fresh Waters – Man and Medium. Oxford: Blackwell Scientific Publications; 1988

[20] West L. Worldwide Water Day:A Billion People Worldwide Lack Safe drinking Water; 2006

[21] Pink DH. Investing in Tomorrow's Liquid Gold; 2006. Available at: http://finance.yahoo.com/columnist/trenddesk/3748.yahoo and http://finance.yahoo.com/columnist/article/trenddesk.3748

[22] EPA USA (Environmental Protection Agency). The National Water Quality Inventory: Report to Congress for the year 2002 Report Cycle A Profile. 2002. Washington, DC. (http://www.epa.gov/305b/2002report/factsheet2002305b.pdf). October 2007. Fact Sheet No. EPA 841-F-03-003

[23] Nakamura K, Tockner K, Amano K. River and wetland restoration: Lessons from Japan. Bioscience. 2006;**56**(5):419-429

[24] Román-Silva DA, Rivera L, Morales T, Avila J, Cortés P. Determination of trace elements in environmental and biological samples using improved sample introduction in flame atomic absorption spectrometry (HHPN-AAS;HHPN-FF-AAS). International Journal of Environmental Analytical Chemistry. 2003;**83**(4):327-341

[25] Naja GM, Volesky B. Toxicity and sources of Pb, Cd, Hg, Cr, As, and radionuclides in the environment. Chapter 2. In: Wang LK, Chen JP, Hung Y, Shammas NK, editors. Heavy Metals in the Environment. Boca Raton: CRC Press; 2009

[26] Andreoni V, Gianfreda L. Bioremediation and monitoring of aromatic-polluted habitats. Applied Microbiology and Biotechnology. 2007;**76**:287-308

[27] Farhadian M, Larroche C, Borghei M, Troquet J, Vachelard A. Bioremediation of BTEX-contaminated groundwater through bioreactors. 4me colloque Franco-Roumain de chimie applique'e, Universite′ Blaise Pascal, Clermont-Ferrand, France, 28 June–2 July 2006. 2006. p. 438

[28] McGuire JT, Long DT, Hyndman DW. Analysis of rechargeinduced geochemical change in a contaminated aquifer. Ground Water. 2005;**43**:518-530

[29] Schreiber ME, Bahr JM. Nitrate-enhanced bioremediation of BTEX-contaminated groundwater: Parameter estimation from natural- gradient tracer experiments. Journal of Contaminant Hydrology. 2002;**55**:29-56

[30] Boopathy R. Factors limiting bioremediation technologies. Bioresource Technology. 2000;**74**:63-67

[31] MacDonald TR, Kitanidis OM, McCarty PL, Roberts PV. Effects of shear detachment on biomass growth and in situ bioremediation. Ground Water. 1999;**37**:555-563

[32] Kampbell T, Don H, Wiedemeier TH, Hansen JE. Intrinsic bioremediation of fuel contamination in ground water at a field site. Journal of Hazardous Maternité. 1996;**49**:197-204

[33] Bala M, Shehu RA, Lawal M. Determination of the level of some heavy. 2008

[34] Ho Y, El-khaiary MI. Metal research trends in the environmental field. In: Wang et al, editors. 2009. Heavy Metals in the Environment. London : CRC Press, Taylor and Francis Group

[35] Long ER, Macdonald DD, Smith SL, Calder FD. Incidence of adverse biological effects within ranges of chemical concentrations in marine and estuarine sediments. Environmental Management. 1995;**19**:81-97

[36] Nriagu JO, Pacyna JM. Quantitative assessment of worldwide contamination of air, water and soils by trace metals. Nature. 1988;**333**:134-139

[37] US EPA. National Recommended Water Quality Criteria. Federal Register, 63; 1998. 68354-68364

[38] WHO (World Health Organisation), 1987, Air Quality Guidelines for Europe, Copenhagen, 1987

[39] Babel S, Kurniawan TA. Low cost adsorbents for heavy metals uptake from contaminated water, a review. Journal of Hazardous Materials. 2003;**97**:219-243

[40] Acar YB, Gale RJ, Alshawabkeh AN, Marks RE, Puppala S, Bricka M, Parker R. Electrokinetic remediation – Basic and technology status. Journal Hazardous of Materials. 1995;**40**:117-137

[41] Ma QY, Traina SJ, Logan TJ, Ryan JA. Effects of aqueous Al, Cd, Cu, Fe (II), Ni, Zn on Pb immobilization by hydroxyapatite. Environmental Science and Technology. 1994;**28**:1219-1228

[42] Pamukcu S, Wittle JK. Electrokinetic removal of selected heavy metals from soil. Environmental Progress. 1992;**11**:241-250

[43] Brown MJ, Lester JN. Metal removal in activated sludge – Role of bacterial extracellular polymers. Water Research. 1979;**13**:817-837

[44] Lasat MM. Phytoextraction of toxic metals: A review of biological mechanisms. Journal of Environmental Quality. 2002;**31**:109-120

[45] Ebbs SD, Lasat MM, Brady DJ, Cornish J, Gordon R, Kochian LV. Phytoextraction of cadmium and zinc from a contaminated soil. Journal of Environmental Quality. 1997;**26**:1424-1430

[46] Dushenkov V, Kumar P, Motto H, Raskin I. Rhizofiltration – The use of plants to remove heavy – Metals from aqueous streams. Environmental Science and Technology. 1995;**29**: 1239-1245

[47] Baker AJM, McGrath SP, Sidoli CMD, Reeves RD. The possibility of in-situ heavy metal decontamination of polluted soils using crops of metal accumulating plants. Resources, Conservation and Recycling. 1994;**11**:41-49

[48] Philippis RD, Micheletti E. Heavy metal removal with exopolysaccharide-Producing Cyanobacteria, Chapter 4. In: Wang LK, Chen JP, Hung Y, Shammas NK, editors. Heavy Metals in the Environment. Boca Raton: CRC Press; 2009

[49] Vymazal J, Švehla J, Kröpfelová L, Chrastný V. Trace metals in *Phragmites australis* and *Phalaris arundinacea* growing in constructed and natural wetlands. Science of the Total Environment. 2007;**380**:154-162

[50] Shammas NK, Wang LK. Treatment and management of metal finishing industry wastes, Chapter 11. In: Wang LK, Chen JP, Hung Y, Shammas NK. ed. Heavy Metals in the Environment. Boca Raton: CRC Press; 2009

[51] Shammas NK. Remediation of metal finishing brown field sites, as chapter 14. In: Wang LK, Chen JP, Hung Y, Shammas NK, editors. Heavy Metals in the Environment. Boca Raton: CRC Press; 2009

[52] Mulligan CN, Yong RN, Gibbs BF, Engineering Geology. 2001;**60**:193

[53] Dushnekov S, Kapulnik Y, Blaylock M, Sorochisky B, Raskin I, Ensley B. Phytoremediation: A novel approach to an old problem, studies in environmental science, global environmental biotechnology, proceedings of the third biennial meeting of the International Society for Environmental. Biotechnology. 1997;**66**:563-572

[54] Chaney RL, Malik M, Li YM, Brown SL, Angle JS, Baker AJM. Phytoremediaation of soil metals. Current Opinion in Biotechnology. 1997;**8**:279-284

[55] Raskin I, Smith RD, Salt DE. Phytoremediaiton of metals: Using plants to remove pollutants from the environment. Current Opinion in Biotechnology. 1997;**8**(2):221-226

[56] Bonanno G, Guidice RL. Heavy metal bioaccumulation by the organs of Phragmites Australis (common reed) and their potential use as contamination indicators, Ecological Indicators. 2010;10:639-645

[57] Zurayk R, Sukkariah B, Baalbaki R. Common hydrophytes as bioindicators of nickel, chromium and cadmium pollution. Water, Air and Soil Pollution. 2001;**127**:373-388

[58] Sawidis T, Chettri MK, Zachariadis GA, Stratis JA. Heavy metals in aquatic plant and sediments from water systems in Macedonia Greece. Ecotoxicology and Environmental Safety. 1995;**32**:73-80

[59] Tilman D. Biodiversity:Population versus ecosystem stability. Ecology. 1996;**77**:350-363

[60] Quan WM, Han JD, Shen AL, Ping XY, Qian PL, Li CJ, Shi LY, Chen YQ. Uptake and distribution of N, P and heavy metals in three dominant salt marsh macrophytes from Yangtze River estuary, China. Marine Environmental Research. 2007;**64**:21-37

[61] US EPA,1997, Recent Developments for In-Situ Treatment of Metal Contaminated Soils, Contract 68-W5-0055

[62] Salt DE. Phytoremediation – A novel strategy for the removal of toxic metals from the environment using plants. Biotechnology. 1995;**13**:468-474

[63] Sela M, Garty J, Tel-Or E. The New Phytologist. 1989;**112**:7

[64] Sela M, Fritz E, Hutterman A, Tel-Or E. Physiologia Plantarum. 1990;**79**:547

[65] Stratford HK, William TH, Leon A. Effects of heavy metals on water hyacinths (*Eichhornia crassipes*). Aquatic Toxicology. 1984;**5**(2):117-128

[66] Demirezen D, Aksoy A. Accumulation of heavy metals in *Typha angustifolia* (L.) and *Potamogeton pectinatus* (L.) living in sultan marsh (Kayseri, Turkey). Chemosphere. 2004;**56**:685-696

[67] Vajpayee P, Tripathi RD, Rai UN, Ali MB, Singh SN. Chemosphere. 2000;**41**:1075

[68] Sarma H. Metal hyper-accumulation in plants: a review focusing on Phytoremediation technology. Journal of Environmental Science and Technology. 2011;**4**(2):118-138

[69] Danh LT, Truong R, Mammucari TT, Foster N. Vetiver grass, vetiveria zizanioides: A choice plant for phytoremediation of heavy metals and organic wastes. International Journal of Phytoremediation. 2009;**11**:664-691

[70] Wenzel WW. Rhizosphere processes and management in plant – Assisted bioremediation (phytoremediation) of soils. Plant and Soil. 2009;**321**:385-408

[71] Zhuang X, Chen J, Shim H, Bai Z. New advances in plant growth promoting Rhizobacteria for bioremediation. Environmental International. 2007;**33**:406-413

[72] Karami A, Shamsuddin ZH. Phytoremediation of heavy metals with several efficiency enhancer methods. African Journal of Biotechnology. 2010;**9**(25):3689-3698

[73] Horsfall M Jr, Abia AA. Sorption of cadmium(II) and zinc(II) ions from aqueous solutions by cassava waste biomass (manihot sculenta cranz). Water Research. 2003;**37**:4913-4923

[74] Papazogloua G, Karantounias GA, Vemmos SN, Bouranis DL. Photosynthesis and growth responses of giant reed to the heavy metals Cd and Ni. Journal of Environmental International. 2005;**31**(1):243-249

[75] www.constructedwetland.co.uk. 2009.Accessed on 9 November 2017

[76] Masters GM. Introduction to Environmental Engineering and Science. Harlow: Prentice Hall; 1991

[77] Zhang T, Ding L, Ren H. Pretreatment of ammonium removal from landfill leachate by chemical precipitation. Journal of Hazardous Materials. 2009;**166**:2-3:911-915

[78] Duggal KN. Elements of environmental engineering. New Delhi: S. Chand Publications; 2008

[79] Vesilind PA, Morgan SM, Heine LG. Introduction to Environmental Engineering, 3/e. Stamford: Cengage Learning; 2010

[80] Kotrba P, Mackova M, Macek T. Microbial Biosorption of Metals. London: Springer; 2011

[81] Mullen MD, Wolf DC, Beveridge TJ, Bailey GW. Sorption of heavy metals by the soil fungi *Aspergillus niger* and *Mucor rouxii*. Soil Biology and Biochemistry. 1992;**24**(2):129-135

[82] Alkorta I, Garbisu C. Phytoremediation of organic contaminants in soils. Bioresources. 2001;**79**:273-276

[83] Salt DE, Smith RD, Raskin I. Phytoremediation. Annual Review of Plant Physiology and Plant Molecular Biology. 1998;**49**:643-668

[84] Baker AJM, Reeves RD, McGrath SP. In situ decontamination of heavy metal polluted soils using crops of metal-accumulating plants—A feasibility study. In: Hinchee RE, Olfenbuttel RF, editors. In-situ Bio-Reclamation. Boston: Butterworth-Heinemann; 1991. 600-605

[85] Kumar P, Dushenkov V, Motto H, Rasakin I. Phytoextraction: The use of plants to remove heavy metals from soils. Environmental Science & Technology. 1995;**29**:1232-1238

[86] Burken JG, Schnoor JL, 1998. Distribution and Volatilisation of Organic Compounds

[87] Burken JG, Schnoor JL. Uptake and metabolism of atrazine by poplar trees. Environmental Science & Technology. 1997;**31**:1399-1406

[88] Smith RAH, Bradshaw AD. Stabilization of toxic mine wastes by the use of tolerant plant populations. Transactions of the Institution of Mininig and Metallurgy Section A. 1972;**81**:230-237

[89] Bañuelos GS, Ajwa HA, Mackey LL, Wu C, Cook S, Akohoue S. Evaluation of different plant species used for phytoremediation of high soil selenium, Journal of Environmental Quality. 1997;**26**:639-646

[90] Susarla S, Medina VF, Mc Cutcheon SC. Phytoremediation: An ecological solution to organic chemical contamination. Ecological Engineering. 2002;**18**:647-658

[91] Schnoor JL, Licht LA, McCutcheon SC, Wolfe NL, Carreira LH. Phytoremediation of organic and nutrient contaminants. Environmental Science & Technology. 1995;**29**:318A-323A

[92] Best EPH, Zappi ME, Fredrickson HL, Sprecher SL, Larson SL, Ochman M. Screening of aquatic and wetland plant species for phytoremediation of explosives contaminated ground water for the Iowa army ammunition plant. Annals of the New York Academy of Sciences. 1997;**829**:179-194

[93] Watanabe ME. Phytoremediation on the brink of commercialization. Environmental Science and Technology. 1997;**31**:182A-186A

[94] Simonich SL, Hites RA. Organic pollutant accumulation in vegetation. Environmental Science & Technology. 1995;**29**:2905-2914

[95] Shimp JF, Tracy JC, Davis LC, Lee E, Huang W, Erickson LE, Schnoor JL. Beneficial effects of plants in the remediation of soil and groundwater contaminated with organic materials. Critical Reviews in Environmental Science and Technology. 1993;**23**:41-77

[96] Pengra BW, Johnston CA, Loveland TR. Mapping an invasive plant, Phragmites Australis, in coastal wetlands using the EO-1 hyperion hyperspectral sensor. Remote Sensing of Environment. 2007;**108**:74-81

[97] Chang Y, Corapcioglu MY. Plant-enhanced subsurface bioremediation of nonvolatile hydrocarbons. Journal of Environmental Engineering. 1998;**112**:162-169

[98] Jones KC. Organic contaminants in the environment, New York: Elsevier Applied Science metals in water collected from two pollution prone irrigation areas around Kanometropolis. Bayero Journal of Pure and Applied Sciences. 1991;**1**(1):36-38

[99] Weis JS, Weis P. Metal uptake, transport and release by wetland plants: Implications for phytoremediation and restoration review. Environment International. 2004;**30**:685-700

[100] Pradhan SP, Conrad JR, Paterek JR, Srivastava VJ. Potential of phytoremediation treatment of PAHs in soil at MGP sites. Journal of Soil Contamination. 1998;**7**:467-480

[101] Bollag JM, Mertz T, Otjen L. Role of micro organisms in soil remediation. In: Anderson TA, Coats JR. eds. Bioremediation through Rhizosphere Technology. ACS Symposium Series 563. American Chemical Society. York, PA: Maple Press; 1994

[102] Pagter M, Bragato C, Malagoli M, Brix H. Osmotic and ionic effects of NaCl and Na2SO4 salinity on Phragmites australis. Aquatic Botany. 2008. doi :10.1016/j.aquabot.2008.05.005

[103] Brix H. Genetic diversity, ecophysiology and growth dynamics of reed (*Phragmites australis*). Aquatic Botany. 1999;**64**:179-184

[104] Tylova E, Steinbachov'a L, Votrubov O, Gloser V. Phenology and autumnal accumulation of N reserves in belowground organs of wetland helophytes Phragmites Australis and Glyceria Maxima affected by nutrient surplus, Environmental and Experimental Botany. 2008;**63**:28-38

[105] Baldantoni D, Ligrone R, Alfani A. Macro- and trace-element concentrations in leaves and roots of Phragmites australis in a volcanic lake in Southern Italy. Journal of Geochemical Exploration. 2008. doi : 10.1016/j.gexplo.2008.06.007

[106] Saltonstall K, Stevenson JC. The effect of nutrients on seedling growth of native and introduced *Phragmites australis*. Aquatic Botany. 2007;**86**:331-336

[107] Unamuno VIR, De Visscher A, Lesage E, Meers E, Leuridan I, Tack FMG. Cu sorption on Phragmites Australis leaf and stem litter : A kinetic study. Chemosphere. 2007;**69**:1136-1143

[108] Tylova E, Steinbachov'a L, Votrubov O, Lorenzen B, Brix H. Different sensitivity of Phragmites Australis and Glyceria Maxima to high availability of ammonium-N. Aquatic Botany. 2008;**88**:93-98

[109] Curn V, Kubatova B, Vavrova P, Krivackova-Sucha O, Cizkova H. Phenotypic and genotypic variation of *Phragmites australis*: Comparison of populations in two human-made lakes of different age and history. 2007;**86**:321-330

[110] Hootsmans MJM, Wiegman F. Four helophyte species growing under saltstress: Their salt of life ? Aquatic Botany. 1998;**62**:81-94

[111] Clevering OA, Lissner J. Taxonomy, chromosome numbers, clonal diversityand population dynamics of *Phragmites australis*. Aquatic Botany. 1999;**64**:185-208

[112] Van der Werff M. Common reed. In: Rozema J, Verkleij JAC, editors. Ecological Responses to environmental Stress. The Netherlands: Kluwer Academic Publishers; 1991. pp. 172-182

[113] Jean L, De M. Constructed wetlands for sludge dewatering, water sci. Technology. 1997;**35**:279-285

[114] Jiang X, Wang C. Zinc distribution and zinc-binding forms in *Phragmites australis* under zinc pollution, Journal of Plant Physiology. 2008;165:697-704

[115] Fediuc E, Erdei L. Physiological and biochemical aspects of cadmium toxicity and protective mechanisms induced in Phragmites Australis and *Typha latifolia*. Journal of Plant Physiology. 2002;**159**:265-271

[116] Rousseau DPL, Vanrolleghem PA, De Pauw N. Constructed wetlands in flanders: A performance analysis. Ecological Engineering. 2004;**23**:151-163

[117] Gessner MO. Breakdown and nutrient dynamics of submerged phragmites shoots in the littoral zone of a temperate hardwater lake. Aquatic Botany. 2001;**66**:9-20

[118] Clemens S, Plamgren MG, Kramer U. A long way ahead: Understanding and engineering plant metal accumulation. Trends in Plant Science. 2002;**7**:309-315

[119] Outten CE, O'Halloran TV. Femtomolar sensitivity of metalloregulatory proteins controlling zinc homeostasis. Science. 2001;**292**:2488-2492

[120] Campbell PGC, Lewis AG, Chapman PM, Crowder AA, Fletcher WK, Imber B. Biologically Available Metals in Sediments; 1988. NRCC No . 27694, Ottawa, Canada

[121] Crowder A. Acidification, metals and macrophytes. Environmental Pollution. 1991;**71**: 171-203

[122] Vymazal J, Brix H, Cooper PF, Green MB, Haberl R. Constructed Wetlands for Wastewater Treatment in Europe. Leiden, The Netherlands: Backhuys Publishers; 1998. pp.366

[123] Bragato C, Brix H, Malagoli M. Accumulation of nutrients and heavy metals in *Phragmites australis* (cav.) trin. Ex steudel and *Bolboschoenus maritimus* (L.) palla in a constructed wetland of the Venice lagoon watershed. Environmental Pollution. 2006;**144**:967-975

[124] Gardea-Torresdey JL. Phytoremediation: Where does it stand and where will it go? Environmental Progress. 2003;**22**(1):A2-A3. Editorial

[125] Bonang Nkoane BM, Gerald Sawula M, Wibetoe G, Lund W. Identification of Cu and Ni indicator plants from mineralised locations in Botswana. Journal of Geochemical Exploration. 2005;**86**(3):130-142

Chapter 3

Biosorption of Heavy Metals by *Candida albicans*

Ismael Acosta Rodríguez,
Juan Fernando Cárdenas-González,
Víctor Manuel Martínez Juárez,
Adriana Rodríguez Pérez,
María de Guadalupe Moctezuma Zarate and
Nancy Cecilia Pacheco Castillo

Additional information is available at the end of the chapter

http://dx.doi.org/10.5772/intechopen.72454

Abstract

The objective of this work was to study the resistance and removal capacity of heavy metals by the yeast *Candida albicans*. The resistance of some heavy metals was analyzed: the yeast grows in 2000 ppm of chromium, zinc, lead, and copper, 1500 ppm of arsenic (III), 500 ppm of silver, and little bit in cobalt (300 ppm) and mercury and cadmium (200 ppm). Analyzing its potential to remove heavy metals, it can efficiently remove is as follows: Cr(VI) (76%), lead (57%), silver (51%), cadmium (46%), fairly arsenic(III) (40% with the modified biomass), cobalt (37%), mercury (36%), copper (31%), little bit zinc (22%), and fluoride (10%). We determine the optimal characteristics for chromium(VI) removal in living cells and death biomass. The ideal conditions for the removal of 50 mg/L of Cr(VI) in living cells were 28°C, pH 7.0, and 10×10^6 yeast/mL, with glycerol-like carbon source. In dead yeast biomass, the ideal conditions for removal of metal are 200 mg/L of Cr(VI), 60°C, pH 1.0, 20 h, and 5 g of biomass.

Keywords: biosorption, heavy metals, *Candida albicans*, bioremediation, microorganism

1. Introduction

Heavy metals are grouped in the category of 53 elements with a specific weight greater than 5 g/cm^3. Some elements like zinc, copper, manganese and cobalt are considered as micronutrients, while others like cadmium, lead, mercury or chromium have no biological functions in plants. Heavy metals do not biodegrade in soil, in which matrix they are as free metal ions, interchangeable metal ions, soluble metal complexes, metals bound in organic materials, precipitates or in insoluble compounds like oxides, carbonates, and hydroxides or also may form part of the silicates [1]. Human-induced pollution can result from mining, industrial, agricultural, military, and nuclear activities, which can induce high concentrations of heavy metals that can enter the food chain and represent a long-term risk potential for the environment and human health [2].

At present, there is a great concern in the world due to the considerable increase in the indices of contamination of industrial effluents by heavy metals such as chromium, nickel, cadmium, lead, and mercury [2]. These toxic substances tend to persist indefinitely in the environment, compromising the well-being and balance not only of the fauna and flora existing in the ecosystem but also the health of people living in the surrounding communities, through their accumulation and entry into the food chain [3]. Among the various effects produced by heavy metals in plants are necrosis at the tips of leaves, inhibition of root growth, and at worst the total death of the plant. In humans, heavy metals can become very toxic when introduced into the organism. At high concentrations, these can cause skin rashes, stomach upset (ulcers), respiratory problems, weakening of the immune system, damage to the kidneys and liver, hypertension, alteration of genetic material, cancer, neurological disorders, and even death [4]. World Health Organization (WHO) established that the maximum concentration of heavy metal ions in water should be in the range of 0.01–1 ppm [5]; however, concentrations of heavy metal ions are up to 450 ppm in effluents [6].

Among the main industrial sectors that are sources of contamination of heavy metals are mining, cement industry, dye industry, tanning, electroplating, steel production, photographic material, corrosion paints, energy production, textile fabrication, wood preservation, aluminum anodizing, water cooling, and others [1, 2]. The environmental impact generated by these toxic substances has led the scientific community to develop different methods for the treatment of industrial effluents contaminated with these substances, which are precipitation, oxidation-reduction, exchange ionic, filtration, electrochemical treatment, membrane technologies, and recovery by evaporation. However, these methods have been quite costly and inefficient, especially when the metal concentration is very low, as well as the formation, disposal, and storage of sludge and wastes, originating during the processes, which become a major problem to solve [7].

Adsorption is the preferential accumulation of a substance on the surface of a normally porous solid. The substance that is adsorbed is called adsorbate and may be an ion or molecule. Furthermore, the solid on which the adsorption occurs is known as an adsorbent [8], and biosorption is a phenomenon widely studied in the bioremediation of sites impacted by pollution. The study of microorganisms and bioadsorbent material is constantly growing with the use of microbial consortia, which would increase yield's uptake of certain specified metals or mixtures thereof [7, 8]. There is evidence of isolation of resistant microorganisms to heavy metals and the use of microbial biomass for the removal of heavy metals from industrial wastewater and/or contaminated water: the isolation and characterization of a variant manganese-resistant strain of *Saccharomyces cerevisiae* [9]; *Pichia guilliermondii* resistant to heavy metal [10]; *Candida albicans* resistant to crude oil [11]; *Pichia anomala, Candida krusei,* and *Cryptococcus laurentii* tolerated high concentrations of zinc (up to 20 mM) [12]; the yeast *S. cerevisiae* partly retains heavy metals (Cu, Fe, Pb, Zn, Ba) and arsenic from soil extracts [2]; hydroxyapatite (HAp)/yeast biomass composites for the removal of Pb^{+2} [13]; removal of zinc by *Pichia kudriavzevii* A16 [14]; copper(II) and phenol adsorption by *Candida tropicalis* [15]; the removal of copper(II) by *Candida krusei* [16]; the application of bifunctional *Mangifera indica* L.–loaded *Saccharomyces cerevisiae* as efficacious biosorbent for bivalent cobalt and nickel cations [17]; the biosorption of Cr(VI) from aqueous solutions by *Candida albicans* and *Cryptococcus neoformans* isolated from leather works [18]; and Cr(VI) reduction in a chromate-resistant strain of *Candida maltosa* [19], with highly satisfactory results. This chapter reports the removal of different heavy metals in an aqueous solution by a strain of *Candida albicans*, which is highly resistant to this metal.

2. Materials and methods

2.1. Microorganism and heavy metal–resistant tests

A yeast strain was isolated from the Bancote River in the Huasteca Potosina (Ciudad Valles, SLP, México) [20], and this was used for the screening. The strain was grown on a petri dish containing modified Lee's minimal medium (LMM) (with 0.25% KH_2PO_4, 0.20% $MgSO_4$, 0.50% $(NH_4)_2SO_4$, 0.50% NaCl, 0.25% glucose, and 2% agar). The pH of the medium was adjusted and maintained at 5.3 with 100 mmol/L of citrate phosphate buffer. The plates were incubated at 28°C for 7 days.

Yeast cultures grown in thioglycolate broth were used as primary inoculums. Heavy metal–resistant tests of the isolated strain, yeast *C. albicans*, were performed on liquid LMM containing the appropriate nutritional requirements and different concentrations of heavy metals (as salt), and the dry weight was determined.

2.2. Identification of yeast

The strain was identified based on its macroscopic characteristics and microscopic observations [21]. Germ tube induction test was performed as follows: 1 × 10^6 yeast/mL is taken, seeded into LMM (added with proline and biotin, 0.5 and 0.001 g/L, respectively), and incubated at 37°C for 3 h. Subsequently, a small sample was taken to analyze in a microscope, the formation of a germinal tube without constriction in its source of origin and with the characteristic shape of hand mirror [21].

Moreover, to examine the formation of chlamydospores, yeast (1 × 10^6 yeast/mL) was grown in corn flour agar medium and incubated at 48–72 h at 28°C, observing under microscope the formation of asexual, thick-walled, and refringent spores, called chlamydospores, which may be intercalated or in terminal position of the hyphae partitions or septate [21].

2.3. Resistance test

Petri dishes were prepared with Sabouraud Dextrose Agar and added with different salts of heavy metals. The prepared plates were inoculated with 1 × 10^6 yeast/mL, uniformly spread throughout the dishes, and incubated at 28°C for 7 days, and the growth of the plates was compared with a control.

2.4. Preparation of biomass

The yeast cells were grown at 28°C in a stirred and aerated liquid media containing thioglycolate broth at a concentration of 8 g/L (w/v). After 7 days of incubation, the cells were recovered by centrifugation (3000 rpm, 10 min) and washed three times in the same conditions with deionized water, and, subsequently, they were dried (80°C, 24 h) in an oven.

Iron oxide–coated biomass was also prepared; 80 mL of 2 M $Fe(NO_3)_3.9H_2O$ was prepared, and 1.0 mL of 10 M NaOH was added to this solution and mixed thoroughly. Twenty grams of the yeast biomass powder was taken in a porcelain pot, and a mixture of iron oxide and NaOH solution was added to the porcelain pot, homogenized, and kept in an oven for 3 h at 80°C. After 3 h, the oven temperature was raised to 110°C and continued for 24 h. The coated biomass powder was separated by crushing with mortar and pestle [22].

2.5. Biosorption tests of chromium(VI) by using dry yeasts

Solutions of Cr(VI) for analysis were prepared by diluting 71.86 mg/L of stock metal solution. The concentration range of Cr(VI) solutions was 50–1000 mg/L. The pH of each solution was adjusted to the required value by adding 1 M H_2SO_4 solution before mixing with the microorganism. The biosorption of the metal by yeast dry cells was determined at different concentrations of 100 mL Cr(VI) solution, with 1 g of yeast biomass, at 100 rpm, and the sample was filtered. The filtrate containing the residual concentration of Cr(VI) was determined spectrophotometrically. For the determination of the rate of metal biosorption, the solution of Cr(VI) was used at concentrations of 200, 400, 600, 800, and 1000 mg/L. The supernatant was analyzed for residual Cr(VI) at different times after a contact period. For the determination of factors such as pH and temperature, seven solutions were analyzed, which included pH 1.0, 2.0, 3.0 and temperatures of 28°C, 40°C, 50°C, and 60°C. Moreover, biosorption to the contaminated earth and water was examined. Four Erlenmeyer flasks containing 5 g of fungal biomass, 20 g of contaminated earth and 20 mL of water (297 mg Cr(VI)/g earth or 155 mg Cr(VI)/L water) from tannery (Celaya, Guanajuato, Mexico), were calibrated to 100 mL with tridesionized water, were incubated during 7 days, stirred at 120 rpm, and filtered in whatman filter paper No. 1. The concentration of Cr (VI) was determined by the 1,5-diphenylcarbazide method.

2.6. Reduction of Cr(VI) by living yeasts

Reduction efficiency of Cr(VI) by living yeasts was examined. To examine the living yeasts, cultures in 100 mL of LMM were inoculated with 1×10^6 yeasts/mL (28°C, 7 days), the cells were centrifuged (3000 rpm, at 4°C, 10 min) and washed three times with sterile trideionized water, and the pellet was resuspended in 3 mL of the same solution and was transferred on a fresh LMM (100 mL with 50 mg/L Cr(VI)). At different times, 1 mL aliquots were removed and centrifuged (3000 rpm, 10 min), and the concentration of Cr(VI) or total Cr in the supernatant was determined [23].

2.7. Removal by different heavy metals by using dry yeasts

Solutions of heavy metals for analysis were prepared by diluting 1 g/mL of stock metal solution. The concentration range of heavy metal solutions was 1–200 mg/L. The pH of each solution was adjusted to the required value by adding 1 M H_2SO_4 solution before mixing with the microorganism. The biosorption of the metals by yeast dry cells was determined at different concentrations of 100 mL heavy metal solution, with 1 g of yeast biomass, at 100 rpm, and the sample was filtered. The concentration of heavy metals was determined in the filtrate with the following methodologies: Cr (VI) spectrophotometrically with dhyphenylcarbazide, Zn, Pb, Hg, Cd, with dithizone, Co by methyl isobutyl ketone, F by specific ion, and Cu, As (III), As (V), Ag by Atomic absorption [23].

3. Results and discussion

3.1. Isolation and identification of a yeast strain tolerant to heavy metals

The microorganism collected from Bancote River was grown on the LMM agar plates containing different concentrations of heavy metals, and the largest colony of yeast was isolated. The colony of the isolated strain grew rapidly within 3–5 days. They were creamy, white-yellowish, glossy, slightly raised colonies with well-defined borders [**Figure 1a**]. Blastoconidia are formed with 3-6 μm in diameter singly in chains or in small loos clusters (**Figure 1b**), and in certain conditions, they form germ tube (**Figure 2**) and chlamydoconidia (**Figure 3**). We designated the strain as *Candida albicans*.

The isolated cells grew in medium on LMM supplemented with different concentrations of heavy metals, about 62.2, 53, and 22% with 2 g/L of Cr(VI) and Pb(II) and 600 mg/L of silver, respectively; growth relative to control (32.3 mg of dry weight without metal) was obtained (**Figure 4**) and, therefore, probably is resistant to the metals. As well, in plate-resistant testing, the yeast grew in 2 g/L of Zn(II), Pb(II), Cu(II), and Cr(VI), 1.5 g/L of As(III), and 200 mg/L of Hg(II) and Cd(II) (**Table 1**). Several microorganisms that are heavy metal resistant have been isolated from different contaminated sites. An important example is a mutant IM3 strain of *S. cerevisiae*, spontaneously grown on solid yeast extract peptone dextrose, medium with highly concentrated Mn (10 mM) [9], *P. guilliermondii* resistant to 400 mM of Mn(II), Zn(II), and Co(II) [10], *C. krusei* and *C. laurentii*, isolated from water, soil, and plant environments, which tolerated high concentrations of Zn(II) (up to 20 mM) [12], *Pichia pastoris* genetically engineered, it grows in 6 mM of $AgNO_3$ (corresponding to 647.2 μg/ml of elemental silver) and 4 mM of SeO_2 (corresponding to 315.8 μg/ml of elemental selenium) [24], *Rhodotorula mucilaginosa* planktonic cells showed the tolerance in the presence of Hg(II) (0.08 mM), Cu(II) (6.40 mM), and Pb(II) (3.51 mM) [25], and the ability of cadmium uptake by metal-resistant yeast, *Candida tropicalis*, isolated from wastewater from industrial area of Sheikhupura, a small town located 40 km central west of Lahore, Pakistan, known for its industry. This yeast grew in 2.5 g/L of Cd(II), 1.4 g/L of Zn(II), 1 g/L of Ni(II), 1.4 g/L of Hg(II), 1 g/L of Cu(II), 1.2 g/L of Cr(VI), and 1 g/L of Pb(II) [26]. The resistance of *Candida albicans* to heavy metals

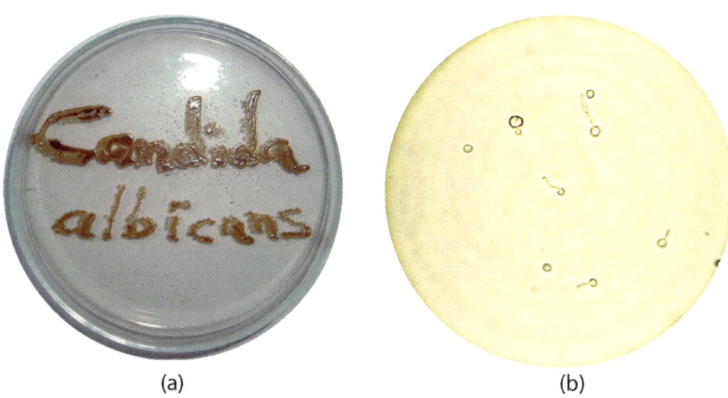

Figure 1. (a) Macroscopic and (b) microscopic morphology of the yeast *Candida albicans*.

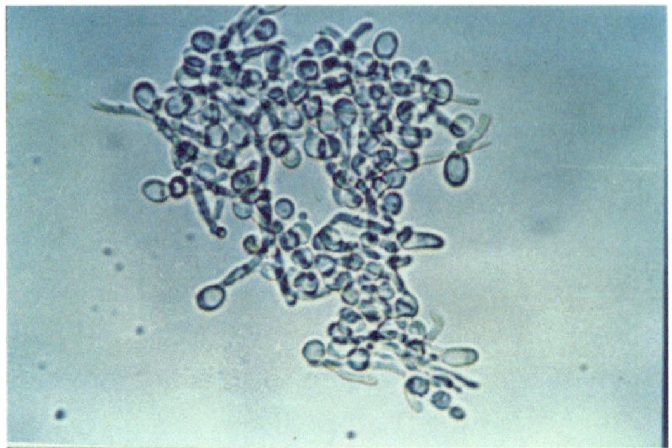

Figure 2. Germ tube induction by *Candida albicans.*

is sufficiently high compared to other yeasts. Therefore, we used the yeast for the following biosorption assays.

3.2. Removal of chromium(VI) by dry cells of *Candida albicans*

3.2.1. *Effect of pH*

The effect of pH on biosorption by using dry cells was examined. First, the capacity of biosorption of heavy metals in dry cells of *Candida albicans* was examined. **Figure 5** shows the effect of incubation time and pH on the biosorption of 50 mg/L of Cr(VI) by the biomass of *A. C. albicans*. It was found that a higher removal, which is proportional to the biosorption, occurs at 2 days and at a pH of 1.0. It was reported a time of 24 h *Cyberlindnera fabianii, Wickerhamomyces anomalus,* and *Candida tropicalis,* at a pH range between 2 and 4 for the three species [27]; the removal of Cr(VI) (100%) by *Cyberlindnera fabianii* at 48 h [28]; *Candida tropicalis* isolated from chromium-contaminated site removal 50 mg/L of the metal at 48 h [29]; and *Candida intermedia* in the biosorption of Cr(III) and Cr(VI) were reported [30]. Permeability and porosity of the cell wall can affect the incubation time of each

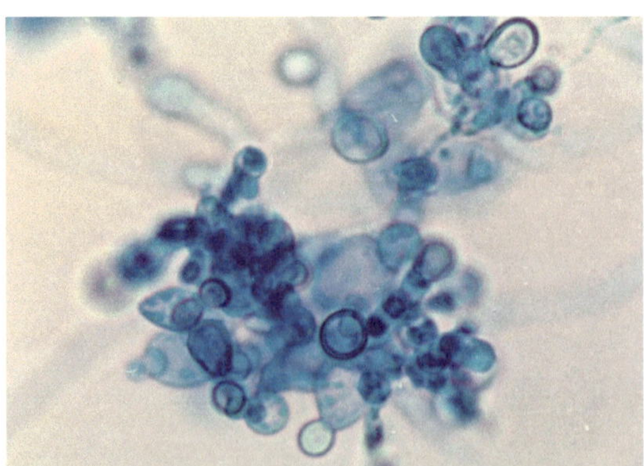

Figure 3. Formation of chlamydoconidia *Candida albicans.*

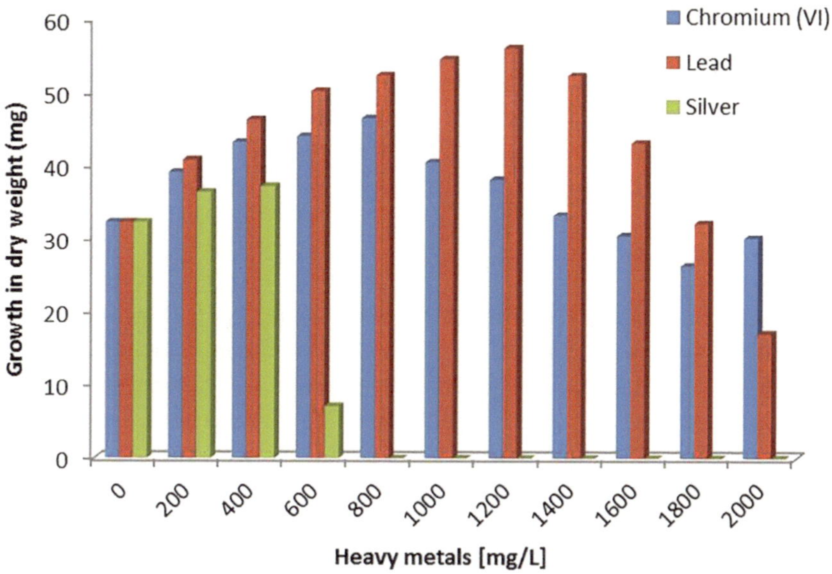

Figure 4. Growth in dry weight of *Candida albicans* with different heavy metal concentrations. 1 × 10^6 yeast/mL, 28°C, 7 days of incubation, 100 rpm.

microorganism, giving greater or lesser exposure of the functional groups in the cell wall of the biomass analyzed [31]. With acid pH used in these experiments, Cr(VI) has negative charge, and the removal by protonation of the biosorbent surface is favored, which induces a strong attraction to these anions, increasing biosorption and, therefore, the removal of the solution. However, if the pH increases, the concentration of OH^- ions increases, and the

Heavy metals	Growth of heavy metal concentration (mg/L)
Zinc	2000
Lead	2000
Copper	2000
Chromium(VI)	2000
Arsenic(III)	1500
Silver	500
Fluor	400
Cobalt	300
Mercury	200
Cadmium	200

Table 1. Growth in LMM in plate of *Candida albicans* with different heavy metals. 1 × 10^6 yeast/mL, 28°C, 7 days of incubation.

positive positions are reduced in the adsorbent surface, releasing the ions of Cr(VI) to the solution. This explains why at higher pH, removal of Cr(VI) [32] decreases, though not always desorption of anions of Cr(VI), is observed with an increase of pH, so biosorption is not the only mechanism occurring, but the reduction of Cr(VI) by organic matter also occurs, due to the high oxidation potential of these species, forming Cr(III), which is an insoluble species in basic medium.

3.2.2. Effect of the temperature

Temperature is also a critical parameter in the removal of Cr(VI) (**Figure 6**); at higher temperatures, we observe greater removal: at 60°C, 100% of the metal is removed in 20 h, and 76% is removed in 24 h at 28°C. These results are similar to those reported for *Pichia jadinii* M9 and *Pichia anomala* M10. More rapidly with an increment in temperature, with an optimum value of 30°C [33], in the removal of copper for encapsulated *Candida krusei* [34], and for *C. neoformans*, with a maximum adsorption capacity for Cr(VI) ion was observed at 28°C [18]. When the temperature increases, the rate of removal of Cr(VI) increases and the contact time required for complete removal of the metal decreases, increasing the redox reaction rate [35].

3.2.3. Effect of the initial concentration of chromium(VI) at 28 and 60°C

The effects of initial concentration of Cr(VI) was examined. At room temperature, the concentration influences the removal of the metal; at lower concentration of the same, removal is faster (at 10 days, 200 ppm, the removal is 100%, and with 1 g/L only, removal is 20% at same time) and at 180 min, 800 and 1000 ppm, are removed, respectively (**Figure 7a**). At 60°C, we observed a removal of 100% at 8 days with 200 mg/L, and only 48% with 1 g/L of the metal

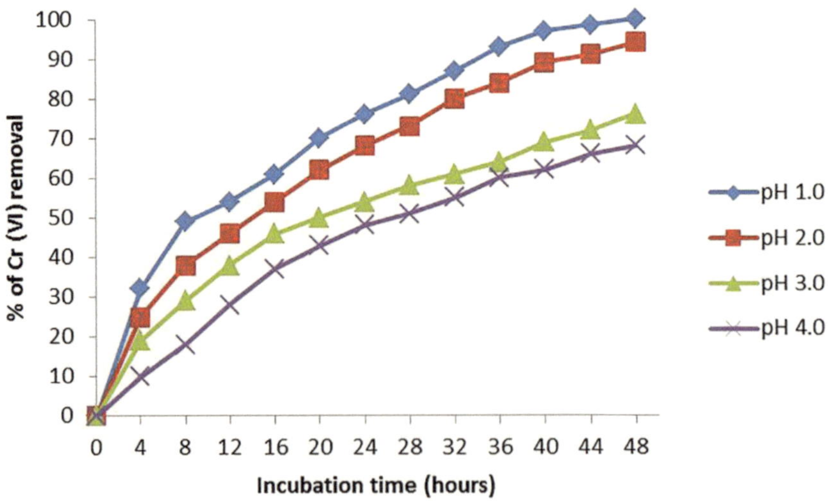

Figure 5. Effect of incubation time and pH on Cr(VI) removal by *C. albicans*. 50 mg/L Cr(VI), 100 rpm, 28°C, and 1.0 g of yeast biomass.

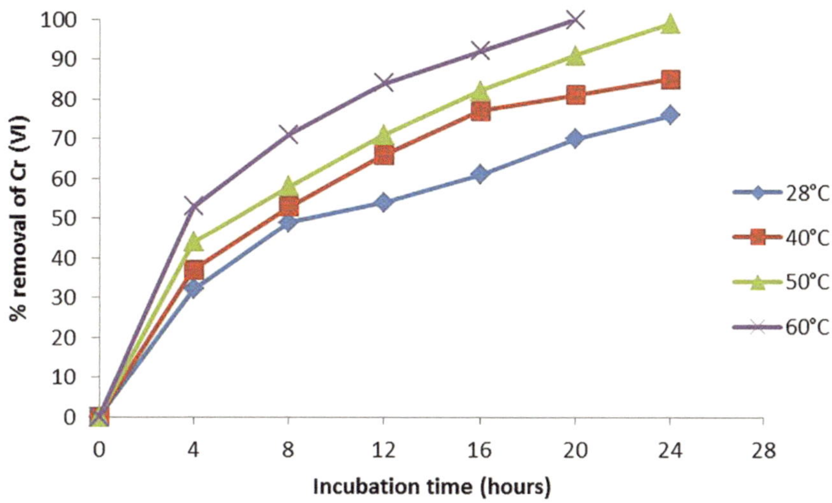

Figure 6. Effect of the temperature on Cr(VI) removal by *C. albicans*. 50 mg/L Cr(VI), 100 rpm, pH 1.0, and 1.0 g of yeast biomass.

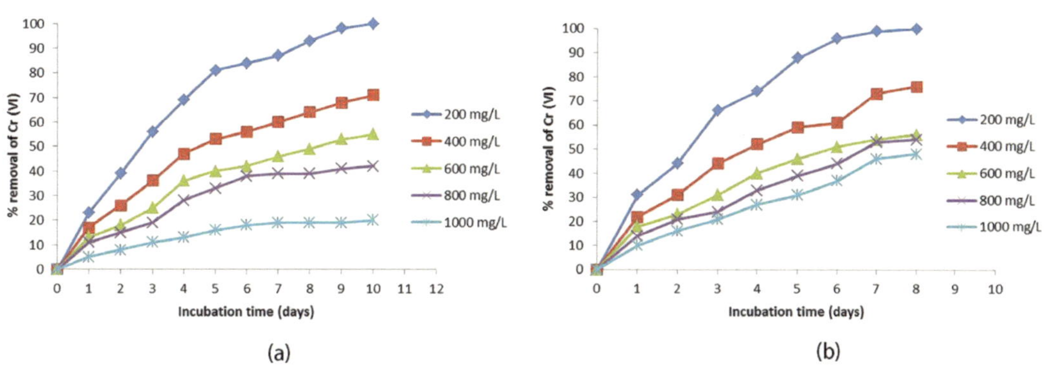

Figure 7. Effect of initial metal concentration on chromium(VI) removal by *C. albicans*. (a) 28°C and (b) 60°C, pH 1.0, 100 rpm, and 1 g of yeast biomass.

(**Figure 7b**), which may be due to sorption happening at low concentrations, but at higher concentrations, possibly when positive positions were saturated, precipitation occurs (which is a slower process) [36]. Some yeasts are not affected by the concentration of the metal, like the three yeast strains isolated from sediments in Morocco [27], and in other yeasts, the removal of metal increases in direct proportion to the increase of the concentration of Cr(VI) in the solution like *Rhodotorula mucilaginosa* isolated from the effluent of Chittaranjan locomotive workshop effluent samples [25], for *Pichia jadinii* M9 and *Pichia anomala* M10 [33], metal-resistant yeast, *Candida tropicalis* [26]. *C. neoformans* showed a higher biosorption capacity at low concentrations of metal ions (0.2 mg/L) [18], *Rhodotorula mucilaginosa* for the removal of copper [37], and *Yarrowia* strains isolated from sediments of mercury-polluted estuarine water [38].

3.2.4. *Effect of the initial concentration of biosorbent*

The effect of the initial concentration of biosorbent on 50 mg/L of Cr(VI) was examined. It was found that the higher the concentration of the latter, the greater and faster the removal of Cr(VI). One gram of biomass with the removal of 76% was observed at 24 h, whereas for 5 g, the removal time was 16 h (**Figure 8**). These observations explain that the amount of added bioadsorbent determines the number of sites available for biosorption load of chrome anions or any metal contaminant [39]. Similar results were reported for the removal of different heavy metals by *S. cerevisiae*, which showed an increase in removal efficiency on increasing biomass from 0.01 to 0.1 g [40], and for the removal of Pb(II) by *S. cerevisiae* CCTCC AY92003 [41]. However, Zn removal rate of the yeast *Pichia kudriavzevii* A16 was not significantly improved when the initial biomass concentration was raised from 0.05 to 1 g/L [14] and is different for *S. cerevisiae* (BCRC23331), which the biosorption capacity of Ni(II) decreasing with increasing adsorbent dose.

3.2.5. *Removal of chromium(VI) in industrial wastes with yeast biomass*

To analyze the possible use and the ability of *C. albicans biomass* to remove chromium(VI) from sediments and effluents, a removal assay was mounted in an aqueous solution in the presence of 5 g biomass, with nonsterile soil contaminated with 297 mg of Cr(VI)g/L and 100 mL of contaminated water with 155 mg of Cr(VI), resuspending the land in trideionized water at 28°C and stirring at 100 rpm. It was observed that after 7 days of incubation, 74 and 69% of Cr(VI) present in the contaminated water and soil were removed, respectively (**Figure 9**). The ability to remove by biomass is equal to or greater than the other biomass that have been studied, *Candida maltose* RR1 [19], *C. tropicalis* was observed to remove 40% Cd (II) from the wastewater after 6 days and was also able to remove 78%

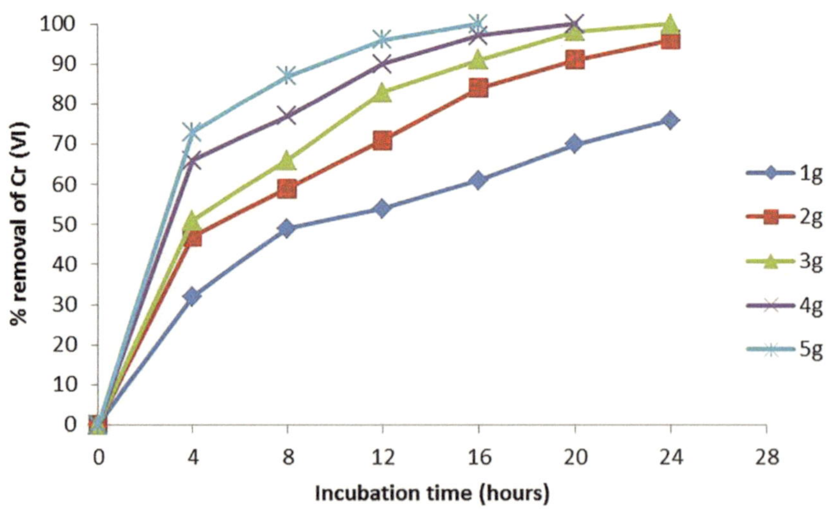

Figure 8. Effect of biomass concentration on chromium(VI) removal by *C. albicans*. 50 mg/L Cr(VI), 28°C, pH 1.0, and 100 rpm.

from the wastewater after 12 days [26], *S. cerevisiae* and *Torulaspora delbrueckii* decrease in 98.1, 83.0, 60.7, 60.5, and 54.2% for turbidity, sulfates, BOD, phosphates and COD, respectively, of the tannery effluent [42], *S. cerevisiae* "wild-type" (WT) parental strain BY4741 very efficient in removing Mn(II), Cu(II), Co (II) from synthetic effluents containing 1–2 mM cations [43].

3.3. Removal of Cr(VI) by the living cells of *Candida albicans*

3.3.1. Effect of incubation time and pH

Next, we will discuss the characteristics of adsorption in the living cells of *C. albicans*. **Figure 10** represents the effect of different pH (4.0, 5.5, and 7.0, maintained at phosphate-citrate buffer 100 mM/L) on the removal of Cr(VI). The speed and rate of metal removal increase as pH goes up. The maximum removal was observed at pH 5.5 and 7.0 (52 and 53% after 7 days of incubation at 28°C and 100 rpm). The ability to remove by living yeast biomass were found were found at pH 4.0 for *C. fabianii* HE650139 and *W. anomalus* HE648168; at pH 3.0 for *C. tropicalis* HE650140, with a percentage removal of 100%, by all living microorganisms [27]; at pH of 5.0–6.5 for the Hg(II) bioremoval by *Yarrowia* strains [38]; at an optimum pH for the strains *P. jadinii* M9 and *P. anomala* M10 of 7.0 and 3.0, respectively, for Cr(VI) reduction [33]; and a pH between 1 and 2 for the removal of Cr(VI) by *Candida utilis* [44]. The decrease of pH causes protonation of the adsorbent surface by attracting ions of Cr(VI) in the solution, so it increases the acidity of the solution, and the biosorption is favored for some microorganisms. As much as the pH increases, the concentration of OH− ions increases too, favoring the presence of Cr(VI) ion valence. This stimulates changes in the biosorbent and prevents metal biosorption [3].

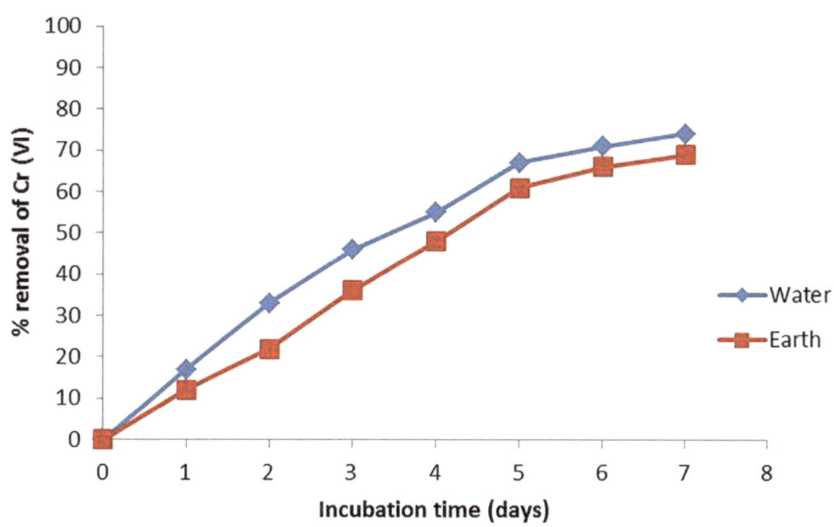

Figure 9. Removal of Cr(VI) from industrial wastes incubated with 5 g of yeast biomass, 100 rpm, 28°C, 20 g, and 100 mL of contaminated soil and water (297 mg Cr(VI)/g earth and 155 mg Cr(VI)/L), respectively.

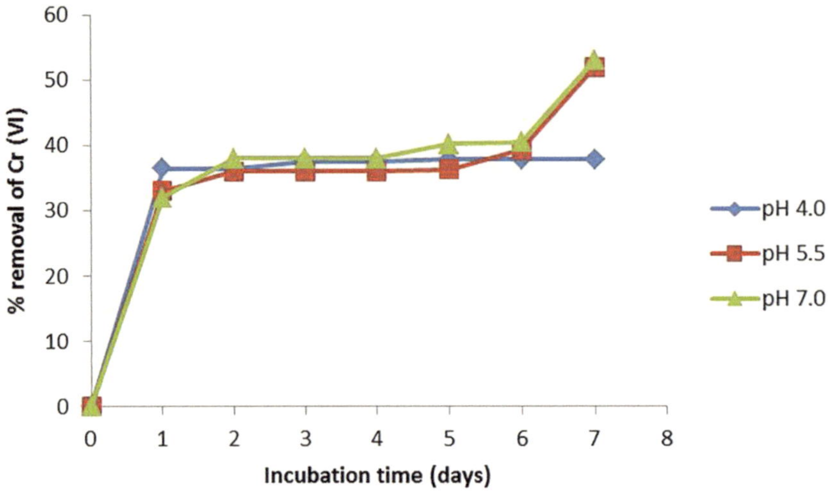

Figure 10. Effect of pH on Cr(VI) removal by biomass of *C. albicans*. 50 mg/L Cr(VI), 100 rpm, and 28°C.

3.3.2. Effect of the initial concentration of the inoculum

The effect of the concentration of cell biomass for the removal capacity of Cr(VI) in the solution (**Figure 11**) was analyzed. At the concentrations tested (1×10^6, 5×10^6, and 10×10^6), the removal capacity was similar (53, 54, and 59%, respectively) for the strains *P. jadinii* M9 and *P. anomala* M10 with pH 7.0 and 3.0, respectively, for Cr(VI) reduction [33] and the removal of Cr(VI) by *Candida utilis* [44]. In contract to our observations, most reports in the literature indicate that the higher amount of biomass increases the percentage of removal: the Hg(II) (6.0, 12.0, 24.0, 48.0, 96.0, and 192.0 mg/L of $HgCl_2$) bioremoval by *Yarrowia* strains [38]. So the greater the amount of the inoculum, the more binding sites for complexing metal (e.g., ions and $HCrO_4^-$ $Cr_2O_7^{2-}$) [1].

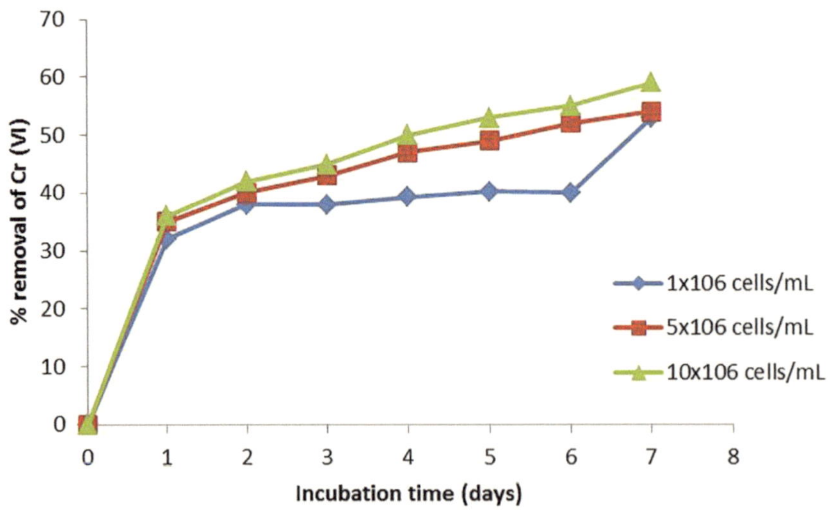

Figure 11. The effect of yeast concentration on the removal of Cr(VI), 50 mg/L Cr(VI), 100 rpm, 28°C, and pH 7.0.

3.3.3. *Effect of the initial concentration of Cr(VI)*

The effect of initial concentration of living cells was examined (**Figure 12**); it was observed that metal removal is more efficient in high concentration of living cells, showing that the removal was 67, 59, 54, and 53% for 200, 150, 100, and 50, mg/L, respectively. The results may be due to the increased amount of ions competing for free functional groups on the surface of the biomass of *C. albicans*. These observations are consistent for the removal of Cr(VI) by *Candida utilis* [44], for contaminated soil for bioremediation of Cr(VI) [45], for the strains *P. jadinii* M9 and *P. anomala* M10 for Cr(VI) reduction (26–104 μg/mL) [33], and for *Candida* sp. isolated from a sewage treatment plant for removal of Cr(VI) [46] and are different from the removal of Cr(VI) by *C. tropicalis* [29].

3.3.4. *Removal capacity of Cr(VI) with different carbon sources*

Moreover, the effect of carbon sources on the removal was examined. In **Figure 13**, we showed the efficiency of the yeast in the removal of Cr(VI) using different carbon sources such as fermentable (glucose, sucrose, and citrate), nonfermentable (succinate), oxidized (glycerol), and commercial (unrefined sugar and brown sugar). We found out that the reduction of Cr(VI) is higher when the medium contains fermentable carbon sources (53% glucose, 97.2% sucrose, and 45% citrate), and removal is high with oxidizable carbon source (68% glycerol), unlike the nonfermentable (42% succinate), while with the other sources of commercial and economic carbon as unrefined (55%) and brown sugar (52%), the removal of Cr(VI) is very similar at 7 days. If we incubate the fungal biomass without a carbon source, there are no changes in the initial Cr(VI) concentration during the experiment (data not shown), suggesting that a carbon source is required to decrease Cr(VI) concentration in the growth medium. Our studies are similar with those reported for *Aspergillus niger* [47], in which the reduction of Cr(VI) is higher when the medium contains fermentable carbon sources (100% glucose, 97.2% sucrose, and 93.35% citrate), and removal is high with oxidizable carbon source (89.9% glycerol), unlike the

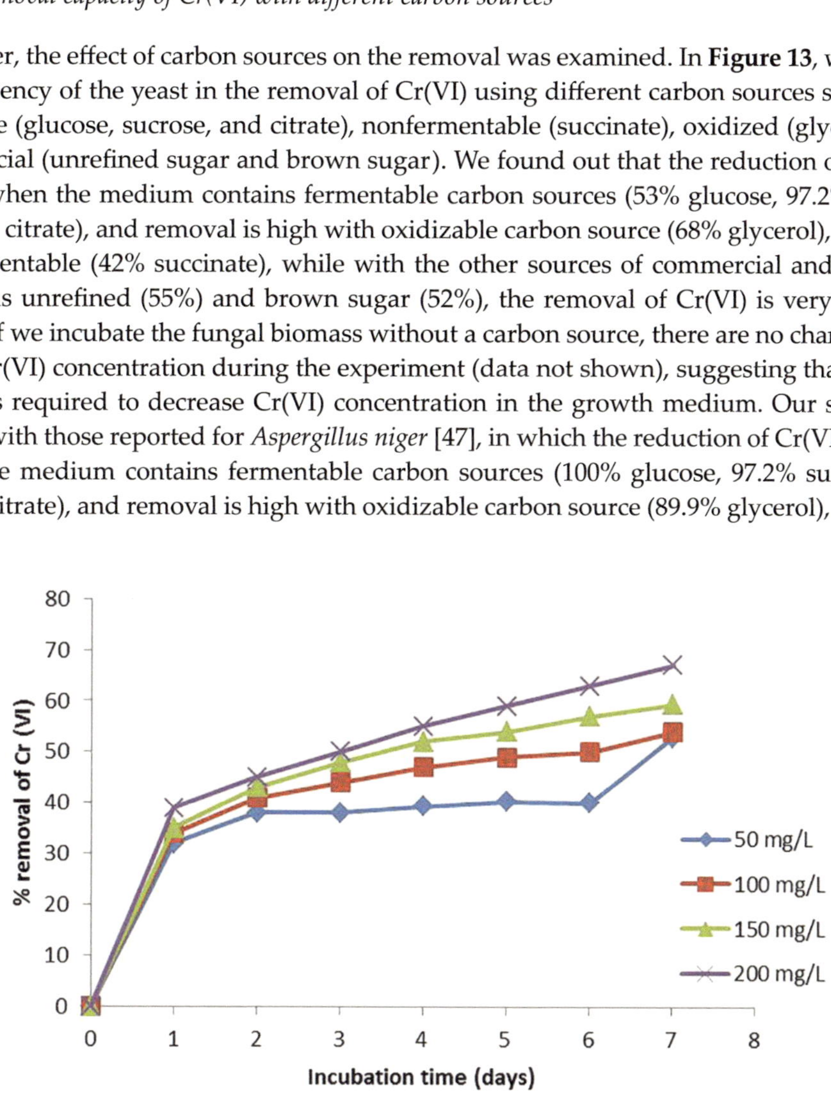

Figure 12. The effect of initial concentration of Cr(VI) on the removal of 50 mg/L Cr(VI), 100 rpm, 28°C, and pH 7.0.

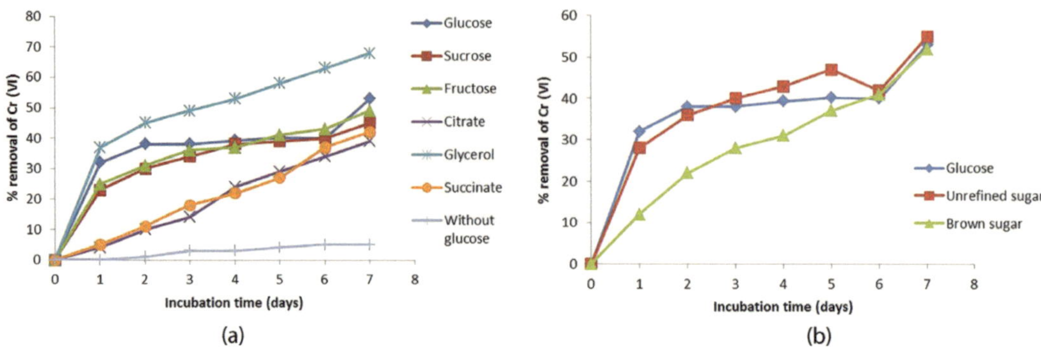

Figure 13. The effect of different carbon sources on the capability of *C. albicans* to decrease Cr(VI) levels in the growth medium, 100 rpm, 28°C, and pH 7.0.

nonfermentable (41.80% succinate), while with the other sources of commercial and economic carbon as unrefined (86.50%) and brown sugar (100%) [47], with *Penicillium lilacinum* using chromate-resistant strains of filamentous fungi indigenously [48], for *Aspergillus foetidus* [49], and the strain Ed8 of *A. niger* chromium resistant, all with glucose as the carbon source [50], but are different from the observations with *Aspergillus* sp., for which the observation was that sodium acetate was the carbon source that induced a greater removal of Cr(VI) [51].

3.3.5. Analysis of the possible use of the yeast C. albicans to the removal of Cr(VI) in nonsterile earth and water contaminated with the same metal

Bioremediation study using soil was also carried out, which was inoculated in 1×10^6 yeast/mL and 100 mL of LMM (pH 5.3) and then 20 g of nonsterile earth and 10 mL of water, contaminated with 50 mg Cr(VI)/g and 50 mg/L of earth and water, respectively, obtained from a factory from the city of Celaya, Guanajuato, Mexico, were added and incubated at 28°C and 100 rpm, observing that after 7 days of incubation, the removal of the metal in the solution was 58 and 62% in earth and water samples, respectively (**Figure 14**), unchanged significantly in the total Cr content. In an experiment conducted in the absence of the yeast, the concentration of Cr(VI) of the samples decreased by about 14% in earth and 6% in water (data not shown), which may be caused by native microflora and reducers present in contaminated samples or components. The removal capacity of Cr(VI) by the fungus is equal or better than that for other yeasts reported such as *C. maltose* RR1 [19], with the removal of Cr(VI) by *C. tropicalis* [29]; the strains *P. jadinii* M9 and *P. anomala* M10, for Cr(VI) reduction [33]; for contaminated soil for bioremediation of Cr (VI) [45]; for *Aspergillus niger* [47]; and different yeasts [52].

3.3.6. Removal of different heavy metals by yeast biomass of C. albicans

Finally, we analyzed the capacity of heavy metal removal by dry cell of the yeast. The results are shown in **Table 2**. The efficiency of yeast removal is shown as follows: Cr(VI) (76%), Pb(II) (57%), Ag(II) (51%), and little bit Zn(II) and F(I) (10%). Both living and dead yeast cells can be effective metal accumulators, and there is evidence that some biomass-based cleanup processes are economically viable [2]. The tolerance of some yeast species to heavy metals, as well

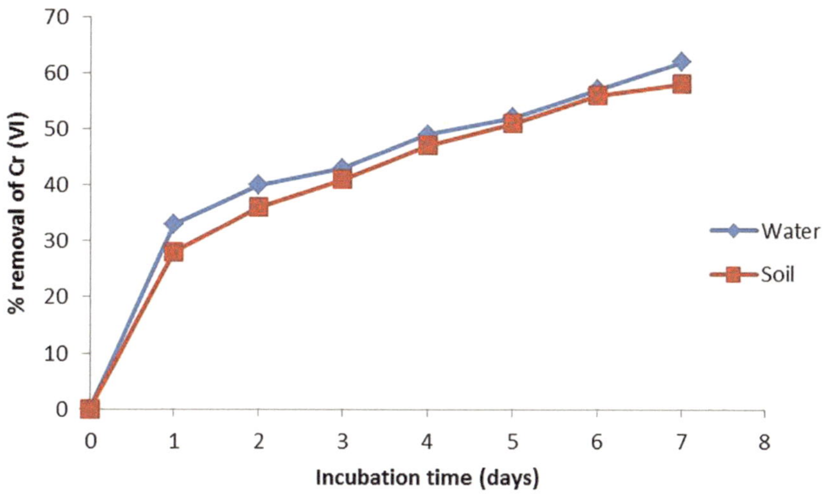

Figure 14. Bioremediation of Cr(VI) from contaminated earth and water (50 mg Cr(VI)/g soil and/or 50 mg/L of water), 28°C, pH 5.3, and 100 rpm.

Heavy metal	pH	Initial concentration (mg/L)	Removal (%)
Chromium(VI)	1.0	50	76
Lead(II)	4.0	100	57
Silver(I)	6.0	100	51
Arsenic(V)*	6.0	1	48
Cadmium(II)	6.0	5	46
Arsenic(III)*	6.0	1	40
Cobalt(II)	4.0	200	37
Mercury(II)	5.5	100	36
Copper(II)	5.0	100	31
Zinc(II)	5.0	100	22
Fluor(I)	6.0	10	10

*Yeast biomass modified with $Fe(NO_3)_3 \cdot 9H_2O$.

Table 2. Removal of different heavy metals by yeast biomass of *Candida albicans*. 28°C, 1 g of yeast biomass, 100 rpm, 24 h.

as the physiological response to them, has also been determined [3]. The removal of heavy metal ions, using yeast as biosorbents, was previously investigated [1–3, 7]. Our results confirm the capacity of the microorganism biomass to remove heavy metals with different effectivities, like yeast-based microbiological decontamination of heavy metal–contaminated soils of Tarnita [2], for bacteria and fungi resistant to crude oil [11], yeasts isolated from water, soil, and plant environments [12], with yeast biomass adsorption for lead (II) [13], removal of zinc by yeast *Pichia kudriavzevii* A16 [14], copper(II) and phenol adsorption by *C. tropicalis* cells in aqueous suspension [15], *C. krusei* for the removal of copper(II) [16], biosorption of cadmium by *C. tropicalis* [26], removal of Cr(VI) by indigenous *Pichia* sp. [33], and Ni(II) biosorption by *S. cerevisiae* [53].

4. Conclusion

We isolated a *C. albicans* yeast, which grew with different heavy metals in LMM and which is probably resistant to the metals. With dead biomass, the removal efficiently 50 mg/L of Cr(VI), 60°C, pH 1.0, 24 h, with 1 g of yeast biomass, with the living cells of *C. albicans*, showed an efficient capacity of reduction (53%) of 50 mg/L Cr(VI) in the growth medium after 7 days of incubation, at 28°C, pH 7.0, 100 rpm and with an inoculum of 10×10^6 yeast/mL, and removal efficiently Cr(VI) (76%), lead (57%), silver (51%), and cadmium (46%). Finally, these results suggest the potential applicability of *C. albicans* for the remediation of Cr(VI) from polluted soils and waters.

Author details

Ismael Acosta Rodríguez[1]*, Juan Fernando Cárdenas-González[2], Víctor Manuel Martínez Juárez[3], Adriana Rodríguez Pérez[2], María de Guadalupe Moctezuma Zarate[1] and Nancy Cecilia Pacheco Castillo[1]

*Address all correspondence to: iacosta@uaslp.mx

1 Chemical Science Faculty, Center of Research and Graduate Studies, Experimental Mycology, Autonomous University of San Luis Potosí, San Luis Potosí, S.L.P, México

2 Multidisciplinary Academic Unit Middle Zone, Autonomous University of San Luis Potosí, Río Verde, S.L.P., México

3 Institute of Agricultural Science, Academic Area of Veterinary Medicine, University Ranch, Autonomous University of Hidalgo's State, Tulancingo de Bravo Hidalgo, México

References

[1] Abbas SH, Ismail IM, Mostafa TM, Sulaymon AH. Biosorption of heavy metals: A review. Journal of chemical Science and Technology. 2014;**3**(4):74-102

[2] Stefanescu R, Butnariu AE, Zamfirache MM, Surleva A, Ciobanui CI, Pintilie O, Drochioiu G. Yeast-based microbiological decontamination of heavy metals contaminated soils of Tarnita. Carpathian Journal of Earth and Environmental Sciences. 2017;**12**(1):153-159

[3] Basu A, Panda SS, Dhal NK. Potential Microbial Diversity in Chromium Mining Areas: A Review. Bulletin of Environment, Pharmacology and Life Sciences. 2015;**4**(8):158-169

[4] Nava Ruíz C, Méndez Armenta M. Efectos neurotóxicos de metales pesados (cadmio, plomo, arsénico y talio). Archivo de Neurociencias Mexicanas. 2011;**16**(3):140-147

[5] WHO. Chromium in Drinking-Water. Geneva, Switzerland; 2003

[6] Srivastava S, Anil Dwivedi K. Biological wastes the tool for biosorption of arsenic. Journal of Bioremediation & Biodegradation. 2015;**7**:323. DOI: 10.4172/2155-6199.1000323

[7] Tejada-Tovar C, Villabona-Ortiz A, Garcés-Jaraba L. Adsorción de metales pesados en aguas residuales usando materiales de origen biológico. Tecno Lógicas. 2015;**18**(34):109-123

[8] Ahemad M. Bacterial mechanisms for Cr(VI) resistance and reduction: An overview and recent advances. Folia Microbiologica. 2014;**59**:321-332

[9] Anh Do T, Sakai T, Kishida M, Furuta M. Isolation and characterization of a variant manganese resistant strain of *Saccharomyces cerevisiae*. Biocontrol Science. 2016;**21**(4): 253-260

[10] Obergozo J, Abanto M, García R, Ramírez P. Identificación molecular de *Pichia guillermondii* aislada de aguas ácidas de minas en el Perú y su resistencia a metales pesados. Revista Peruana de Biología. 2008;**15**(1):91-95

[11] Moctezuma Zárate MG, Cárdenas González JF, Rodríguez Pérez AS, Enríquez Domínguez E, Tovar Oviedo J, Martínez Juárez VM. Acosta Rodríguez: Isolation and identification of bacteria and fungi resistant to crude oil. Journal of Multidisciplinary Engineering Science and Technology. 2016;**3**(7):5273-5278

[12] Vadkertiova R, Slavikova E. Metal tolerance of yeasts isolated from water, soil and plant environments. Journal of Basic Microbiology. 2006;**46**:145-152

[13] Zhang W, Wanga F, Wang P, Lin L, Zhao Y, Zou P, Zhao M, Chen H, Liu Y, Zhang Y. Facile synthesis of hydroxyapatite/yeast biomass composites and their adsorption behaviors for lead (II). Journal of Colloid and Interface Science. 2016;**477**:181-190

[14] Li CH, Yu J, Wang D, Li L, Yang X, Mab H, Xu Y. Efficient removal of zinc by multi-stress-tolerant yeast *Pichia kudriavzevii* A16. Bioresource Technology. 2016;**201**:43-49

[15] Honfi K, Tálos K, Kőnig-Péter A, Kilár F, Pernyeszi T. Copper(II) and phenol adsorption by cell surface treated *Candida tropicalis* cells in aqueous suspension. Water Air Soil Pollution. 2016;**227**(61):1-14

[16] Luk ChHJ, Yip J, Yuen ChWM, Kwong Pang S, Hung Lam K, Wai Kan Ch. Biosorption performance of encapsulated *Candida krusei* for the removal of copper(II). Scientific Reports. 2017;**7**:2159. DOI:10.1038/s41598-017-02350-7 1

[17] Akout AA, Medhat A, Shaker MA, Albishrid HA. Application of bifunctional *Mangifera indica* L.-loaded *Saccharomyces cerevisiae,* as efficacious biosorbent for bivalent cobalt and nickel cations from different wastewaters: Equilibrium and kinetic studies. Desalination and Water Treatment. 2016;**57**:8967-8980

[18] Acosta I, Rodríguez X, Gutiérrez C, Moctezuma MG. Biosorption of chromium (VI) from aqueous solutions by fungal biomass. Bioinorganic Chemistry and Applications. 2004;**2**(1-2):1-7

[19] Ramırez-Ramırez R, Calvo-Mendez C, Avila-Rodrıguez M, Lappe P, Ulloa M, Vazquez-Juarez R, Gutierrez-Corona JF. Cr(VI) reduction in a chromate-resistant strain of *Candida maltosa* isolated from leather industry. Antonie Van Leeuwenhoek. 2004;**85**:63-68

[20] Acosta-Rodríguez I, Moctezuma-Zárate MG, Tovar-Oviedo J, Cárdenas-González JF. Aislamiento e Identificación de Bacterias y Levaduras Resistentes a Petróleo. Revista Información Tecnológica de Chile. 2011;**22**(6):103-110

[21] López Martínez R, Méndez Tovar LJ, Hernández Hernández F, Castañon Olivares R. Hongos contaminantes comunes en el Laboratorio. In: En Micología Médica. Procedimientos para el diagnóstico de Laboratorio. 2ª ed. México: Trillas; 2004. pp. 99-107

[22] Pokhrel D, Viraraghavan T. Arsenic removal from an aqueous solution by a modified fungal biomass. Water Research. 2006;**40**:549-552

[23] Greenberg AE, Clesceri LS, Eaton AD. Standard Methods for the Examination of Water and Wastewater. 18th ed. Washington, DC: American Public Health Association; 1992. pp. 3.58-3.60

[24] Elahian F, Reiisi S, Shahidi A, Abbas Mirzaei S. High-throughput bioaccumulation, biotransformation, and production of silver and selenium nanoparticles using genetically engineered *Pichia pastoris*. Nanomedicine: Nanotechnology, Biology, and Medicine. 2017;**13**:853-861

[25] Grujić S, Vasić S, Radojević I, Čomić L, Ostojić A. Comparison of the *Rhodotorula mucilaginosa* biofilm and planktonic culture on heavy metal susceptibility and removal potential. Water Air Soil Pollution. 2017;**228**(73):1-8

[26] Rehman A, Anjum MS. Multiple metal tolerance and biosorption of cadmium by *Candida tropicalis* isolated from industrial effluents: Glutathione as detoxifying agent. Environmental Monitoring and Assessment. 2011;**174**:585-595

[27] Bahafid W, Joutey NT, Sayel H, Iraqui-Houssaini M, El Ghachtouli N. Chromium adsorption by three yeas strains isolated from sediments in Morocco. Geomicrobiology Journal. 2013;**30**:422-429

[28] Bahafid W, Joutey NT, Sayel H, Ghachtouli N. Mechanism of hexavalent chromium resistance and removal by microorganisms. Reviews of Environmental Contamination and Toxicology. 2015;**233**:45-69

[29] Bahafid W, Joutey NT, Sayel H, Boularab I, Ghachtouli N. Bioaugmentation of chromium-polluted soil microcosms with *Candida tropicalis* diminishes phytoavailable chromium. Journal of Applied Microbiology. 2013;**115**:727-734

[30] Pas M, Mila R, Draslar K, Pollak N, Raspor P. 2004. Uptake of chromium(III) and chromium(VI) compounds in the yeast cell structure. Biometals. 2004;**17**:25-33

[31] Park D, Yun YS, Jo JH, Park JM. Mechanism of hexavalent chromium removal by dead fungal biomass of *Aspergillus niger*. Water Research. 2005;**39**:533-540

[32] Bai RS, Abraham TE. 2001. Biosorption of chromium(VI) from aqueous solution by *Rhizopus nigricans*. Bioresource Technology. 2001;**79**:73-81

[33] Fernandez PM, Martorell MM, Fariña JI, Figueroa LIC: Removal efficiency of Cr^{6+} by indigenous *Pichia* sp. isolated from textile factory effluent. The ScientificWorld Journal. 2012, Article ID 708213, 6 pages doi:10.1100/2012/708213

[34] Luk CHJ, Yip J, ChW MY, Kwong Pang S, Kim Hung Lam KH, ChW K. Biosorption performance of encapsulated *Candida krusei* for the removal of Copper(II). Scientific Reports. 2017;**7**:2159. DOI: 10.1038/s41598-017-02350-7

[35] Agarwal GS, Kumar H, Chaudari S. 2006. Biosorption of aqueous chromium(VI) by *Tamarindus indica* seeds. Bioresource Technology. 2006;**97**:949-956

[36] Leyva-Ramos R. Fundamentos de adsorción en sistemas liquido-sólido. Situacion del arsenico en la Region Iberica e Iberoamericana. Posibles acciones articuladas e integradas para el abatimiento del arsenico en zonas aisladas. Argentina: CYTED; 2010 Chapter 3, 43 p

[37] Salvadori MR, Ando RA, Oller do Nascimento CA, Benedito Correa B. Intracellular biosynthesis and removal of copper nanoparticles by dead biomass of yeast isolated from the wastewater of a mine in the brazilian Amazonia. Plos One. www.plosone.org. 2014;**9**:1-9

[38] Oyetibo GO, Ishola ST, Ikeda-Ohtsubo W, Miyauchi K, Ilori MO, Endo G. Mercury bioremoval by Yarrowia strain isolated from sediments of mercury-polluted estuarine water. Applied Microbiology and Biotechnology. 2015;**99**(8):3651-3657

[39] Kratochvil D, Volesky B. Advances in the biosorption of heavy metals. Reviews. Tibtech. 1998;**16**:291-300

[40] Salah N, Farhan SN, Khadom AA. Biosorption of heavy metals from aqueous solutions by *Saccharomyces Cerevisiae*. International Journal of Indian Chemical. 2015;**6**:119-130

[41] Xina S, Zenga Z, Zhouc X, Luod W, Shia X, Wange Q, Denga H, Dua Y. Recyclable *Saccharomyces cerevisiae* loaded nanofibrous mats with sandwich structure constructing via bio-electrospraying for heavy metal removal. Journal of Hazardous Materials. 2017;**324**:365-372

[42] Okoduwa SIR, Igiri B, Blessing Udeh C, Edenta C, Gauje B. Tannery effluent treatment by yeast species isolates from watermelon. Toxics. 2017;**5**:6. DOI: 10.3390/toxics5010006. 1-10

[43] Ruta L, Paraschivescu C, Matache M, Avramescu S, Farcasanu IC. Removing heavy metals from synthetic effluents using "kamikaze" *Saccharomyces cerevisiae* cells. Applied Microbiology and Biotechnology. 2010;**85**:763-771

[44] Muter O, Patmalnieks A, Rapoport A. Interrelations of the yeast *Candida utilis* and Cr (VI): Metal reduction and its distribution in the cell and medium. Process Biochemistry. 2001;**36**:963-970

[45] Krishna KR, Philip L. Bioremediation of CR(VI) in contaminated soils. Journal of Hazardous Materials. 2005;**B121**:109-117

[46] Baldi F, Vaughan AM, Olson GJ. Chromium (VI)-resistant yeast isolated fron a sewage treatment plant receiving tannery wastes. Applied and Environmental Microbiology. 1990;**56**(4):913-918

[47] Acosta-Rodríguez I, Cárdenas-González JF, Moctezuma-Zárate MG, Rodríguez Pérez A, Martínez-Juárez VM. Hexavalent chromium (VI) removal by *Aspergillus niger*. Surajis Das, ed. In: Meta-microbe interactions and bioremediation: Principles and applications for toxic metals. CRS Press Book. Taylor Francis. 2017. Chapter 42, pp. 673-688. ISBN: 9781498762434 (ebook)

[48] Acevedo-Aguilar F, Espino-Saldana A, Leon-Rodriguez I, Rivera-Cano ME, Avila-Rodriguez M, Wrobel K, Lappe P, Ulloa M, Gutierrez-Corona JF. Hexavalent chromium removal in vitro and from industrial wastes, using chromate-resistant strains of filamentous fungi indigenous to contaminated wastes. Canadian Journal of Microbiology. 2006;**52**:809-815

[49] Prasenjit B, Sumathi S. Uptake of chromium by *Aspergillus foetidus*. Journal of Material Cycles and Waste Management. 2005;**7**:88-92

[50] Coreño-Alonso A, Sole A, Diestra E, Esteve I, Gutierrez-Corona JF, Reyna-Lopez GE, Fernandez FJ, Tomasini A. 2014. Cr(VI) mechanisms of interaction of chromium with *Aspergillus niger* var. *tubingensis* strain Ed8. Bioresource Technology. 2014;**158**:188-192

[51] Srivastava S, Thakur IS. Isolation and process parameter optimization of *Aspergillus* sp. for removal of chromium from tannery effluent. Bioresource Technology. 2014;**97**:1167-1173

[52] Menezes LB, Braz VN, Cordeiro CA. Chromium uptake from aqueous effluents by immobilized Baker's yeast. Revista de Microbiología. 1998;**29**(4):1-5

[53] Nguyen ML, Ruey-Shin J. Modification of cross linked chitosan beads with histidine and *Saccharomyces cerevisiae* for enhanced Ni(II) biosorption. Journal of the Taiwan Institute of Chemical Engineers. 2015;**56**:96-102

Recent Trend on Bioremediation of Polluted Salty Soils and Waters Using Haloarchaea

Sonia Aracil-Gisbert, Javier Torregrosa-Crespo and
Rosa María Martínez-Espinosa

Additional information is available at the end of the chapter

http://dx.doi.org/10.5772/intechopen.70802

Abstract

Pollution of soils, sediments, and groundwater is a matter of concern at global level. Industrial waste effluents have damaged several environments; thus, pollutant removal has become a priority worldwide. Currently, bioremediation has emerged as an effective solution for these problems, and, indeed, the use of haloarchaea in bioremediation has been tested successfully. A bibliographic review is here presented to show the recent advances in bioremediation of polluted soil and wastewater using haloarchaea. Several aspects related to the publications in the context of bioremediation and the innovative proposal of using haloarchaea are also analyzed. The results obtained claim that most of the countries show an alarming contamination issue, which focuses the finance into research about environmental friendly remediation approaches to solve this problem. Concerning bioremediation, strategies to treat soils and polluted waters have been much more studied than bioremediation processes addressed to industrial residues. Publications about bioremediation mainly comes from USA, China, and India at the time of writing this work. However, works using haloarchaea in bioremediation came from Kuwait and Spain. Haloarchaea have been investigated as a potential tool for industrial and environmental purposes. Further research is needed to elucidate the optimal growth conditions and environmental parameters for this proposal.

Keywords: bioremediation, haloarchaea, *Haloferax mediterranei*, heavy metals, green biotechnology, nitrogen, (per)chlorates, phosphorous

1. Introduction

Last decades highlight the importance of the increasing water demand, the environmental pollution or the presence of excess nitrogen in the environment, causing serious alterations in soil, water, and atmosphere [1, 2]. The presence of heavy metals in the environment as well as the C:N ratio has been a major problem in soils and waters due to their toxicity; therefore, their elimination before being released into the media is relevant from environmental, economic, and social points of view [3]. There are many methods to remove pollutants, involving physical, chemical, or biological approaches. During recent decades, the latter has received an increasing attention as a way of removing or neutralizing pollutants from a contaminated site using native or introduced microorganisms [4].

One of those biological strategies recently applied is the bioremediation, which is defined as the application of the metabolic capabilities of bacteria, fungi, yeast, algae, plants, and microbial mats to degrade harmful contaminants to living organisms or transform them into less toxic compounds by naturally enhancing degradation processes [3, 5]. Two types of bioremediation might be found according to the way of use: (i) intrinsic *in situ* bioremediation: natural degradation of remaining contaminants in the environment, without any external intervention; (ii) engineered *in situ* bioremediation: requires an external supply to increase the biodegradation rate by accelerating the growth of further organisms through the secretion of essential nutrients. In most of the situations, engineered bioremediation prevails over the intrinsic one except for those cases in which the natural occurrence of biodegradation is faster than the migration [6].

Engineered bioremediation includes two approaches: biostimulation and bioaugmentation. The addition of limiting nutrients to a contaminated environment where all the necessary natural microorganisms are assumed to be present refers to biostimulation. Bioaugmentation controls the quantity of microorganisms with the desired catalytic capability needed for biodegradation of specific compounds [7, 8]. A combination between these disciplines is suggested to be a promising strategy to speedup bioremediation [5]. Nowadays, biological remediation is considered one of the best options for the treatment of contaminated environments [3] due to their numerous advantages such as *in situ* treatment of soils or waters, high efficiency, and no secondary pollution [5, 9].

1.1. Microorganisms and bioremediation: the case of haloarchaea

Microorganisms are the ones which oversee the bioremediation process, thanks to their ability of using environmental contaminants as source of carbon and energy and their small contact surface [6]. Prokaryotes are the main biotechnological agents used for biodegradation of organic matter, although fungi, algae, and protozoa can be used for the same purpose but affecting differently the nature of the compound for the degradation [10]. In addition, phytoremediation has been for a long time as a promising technique to degrade contaminants from soil sediment, surface, and groundwater [11, 12]. It was proposed the use of microbial consortium, such as phosphate solubilizing microbes, in association with phytoremediation to achieve a most trustworthy approach for the enhancement of the remediation of metal-contaminated soils [13].

More recently, some research has shown that archaea can be of high interest for biotechnological purposes including bioremediation. Archaea were first characterized as a group of single-celled prokaryotic microorganisms living in extremophile environments with elevated temperatures (thermophiles), low or high pH (acidophiles, alkalophiles), high salinity (halophiles), or strict anoxia [14]. The ability of those species to grow under a wide range of extreme conditions equally to their mechanisms supporting genetic plasticity [15] make them good candidates for research in different fields, including biotechnology and, consequently, bioremediation [9, 14, 16].

Nowadays, there is an increasing interest in the optimization of bioremediation approaches in high salt environments, which are mostly influenced by the discharge of industrial effluents [2, 17]. The increase of salinity and nitrogen compounds in soils and ground waters during the last years is focusing attention in the physiological and molecular mechanisms involved in salt-stress tolerance and nitrate metabolism [18]. Conventional microbiological processes are not capable of being executed at high salt concentrations, indicating that microbial bioremediation of hypersaline produced water requires halophiles [19, 20].

In this sense, haloarchaea have been successfully tested for biotechnological applications throughout the last decade [16, 21, 22]. This is because of the extreme properties of their enzymes: high thermostability and resistance to denaturing agents such as detergents, organic solvents, and extreme pH [23]. Moreover, their proteins are rich in acidic amino acids, which allow the maintenance of the stable conformation and activity at high salt concentrations [24].

Most of the species from *Halobacteriaceae* and *Haloferacaceae* families can grow under anaerobic conditions, and they are more abundant and diverse than other microbes in environments containing increasing salt concentrations (more than 20–25% salts) [25, 26]. They are found in natural salty lakes and saltmarshes created by human beings, which are extremely alkalinized places, likewise in submarine brine pools and brine pockets within sea ice [23, 27].

General biochemical pathways related to nitrogen cycle, heavy metals, hydrocarbons, or aromatic compounds have been described in haloarchaea as efficient pathways. Consequently, they might be applied for bioremediation proposals in saline and hypersaline wastewater treatments, thanks to their high tolerance to salt, metals, and organic pollutants [3, 14, 25].

1.2. *Haloferax mediterranei*: a good model for viable bioremediation applications

Hfx. mediterranei is a haloarchaeon from the family *Haloferacaceae* [28], first isolated from seawater evaporation ponds near Alicante, Spain [29]. It can grow in a broad range of NaCl concentrations (1.0–5.2 M) [14], thanks to its remarkable metabolic efficiency and genome stability at high salt concentrations [25, 30].

Hfx. mediterranei is also a denitrifying haloarchaeon based on its ability to reduce nitrate and nitrite under oxic and anoxic conditions [31, 32]. In the presence of oxygen, this haloarchaeon takes profit of nitrate and nitrite as nitrogen sources for its own proliferation [33] through assimilation pathway. Under oxygen depletion, it uses both ions as terminal electron acceptors in the respiratory chain via denitrification [32, 34]. Once nitrogenous compounds are consumed, brines become free of nitrogenous compounds such as nitrate, nitrite, or ammonium.

Owing to this specific metabolism, it was demonstrated that *Hfx. mediterranei* was capable of removing most of the nitrogen compounds present in a treated medium, specifically in anoxic conditions after the induction of the denitrification pathway [14]. Its resistance to very high nitrate (up to 2 M) and nitrite (up to 50 mM) concentrations makes them the highest described from a prokaryotic microorganism [3, 35]. Furthermore, the respiratory nitrate reductase of this organism can reduce efficiently bromate and (per)chlorate thanks to the use of these compounds as terminal electron acceptors [36].

Because of its capability of growing faster than most of other members of the same family and their described characteristics, *Hfx. mediterranei* has served as a good model for haloarchaeal physiology and metabolic studies during several decades [30, 37]. Therefore, it represents a highly interesting model for research in potential development of bioremediation approaches in polluted soils, sediments, and waters.

We had reviewed on bioremediation of polluted salty soils and water using haloarchaea in our previous manuscript [3, 22, 25, 34, 36] and had introduced the importance of haloarchaea as a bioremediation tool. Thus, in this manuscript, we statistically investigated the recent trend of the study and estimated the characteristics.

2. Research method

The starting point for this work was the bibliographic-bibliometric review about the most interesting keywords related to the topic. To optimize the search of information, it was used "PubMed" (https://www.ncbi.nlm.nih.gov/pubmed/) as a free search engine, as well as two well-known subscription-based scientific databases: "Scopus" (https://www.scopus.com/) and "Web of Science (WoS)" (https://www.webofknowledge.com/). The main keywords that have been used include "archaea," "bioremediation," "contaminated soil," "haloarchaea," "industrial waste," "salty soil," and "wastewater." In addition, some keywords were employed for the sole purpose to corroborate the total number of publications considering synonyms or similar words as the previous ones, such as "halophilic archaea," "industrial raw," and "salted soil."

The combination between these keywords was used to optimize the search for publications. All the keywords were grouped into two subclasses: the first one includes the terms "archaea," "bioremediation," "haloarchaea," and "halophilic archaea"; the second class involves "contaminated soil," "industrial raw," "industrial waste," "salted soil," "salty soil," and "wastewater." These classes of keywords and their synonyms were crossed, obtaining a total pool of 27 concepts: "archaea and contaminated soil," "archaea and bioremediation," "archaea and industrial raw," "archaea and industrial waste," "archaea and salted soil," "archaea and salty soil," "archaea and wastewater," "bioremediation and contaminated soil," "bioremediation and haloarchaea," "bioremediation and halophilic archaea," "bioremediation and industrial raw," "bioremediation and industrial waste," "bioremediation and salted soil," "bioremediation and salty soil," "bioremediation and wastewater," "haloarchaea and contaminated soil," "haloarchaea and industrial raw," "haloarchaea and industrial waste," "haloarchaea and salted soil," "haloarchaea and salty soil," "haloarchaea and wastewater," "halophilic archaea and contaminated soil," "halophilic archaea and industrial raw," "halophilic archaea and industrial waste," "halophilic archaea and salted soil," "halophilic archaea and salty soil," and "halophilic archaea and wastewater." The following fields were selected in the databases to delimit the results: "article title," "abstract," and "keywords."

In April 2017, the search about number of publications per year was finished, using a range of 27 years (1990–2016). They were obtained regarding the following items: country, type of document, language, and publication date. The language and type of document were not indispensable for analyzing purposes. The research question was: what is the impact of haloarchaeal applications in bioremediation in research/science?

To assess the differences in the search of publications between the three mentioned databases, the analysis of variance (ANOVA) was performed using R-project statistics. ANOVA statistics was chosen, since it is usually employed for this kind of studies. The significant threshold for these analyses had a p value of 0.05 except for those cases where the transformation of the data was not efficient, and it was needed to change it ($p = 0.01$). In those cases where at least one of the levels was statistically different, it was performed a test to delimit the class which is different from the rest or differences between classes. The test used was Tukey's HSD. For the average calculations and graphical representations, Microsoft Excel 2016 was employed.

3. Results

3.1. Analysis of databases results regarding publications per year

First, a contrast between PubMed, Scopus, and Web of Science was performed regarding the number of publications per year. The purpose of this approach was to study if there were significant differences in the search concerning the database based on the keywords selected. To summarize the data, only publications from a range of years (1990–2016) were used in the statistical test of the total 10 keywords, including synonyms. The year 2017 was not considered in this interval due to the inaccuracy of the results and the unpredictable number of total publications at the end of the year.

An ANOVA test was performed to analyze the differences between the three databases (PubMed, Scopus, and Web of Science) and to compare them looking for statistical significance in the number of publications. The requirements for ANOVA test were corroborated as follows: independence of the data, normality of the results, and homogeneity of the variances. ANOVA statistics was done obtaining that at least one of the databases was different ($p < 0.05$), so Tukey's test was performed to distinguish the variable class by comparing each level with the rest (**Figure 1**). Concerning the p value of the results, there are only significant differences between Web of Science and PubMed as it is displayed in **Figure 1** by means of lines referring to similarities. The differences between the databases are remarked in the way PubMed is significantly different from Web of Science ($p = 0.0018$) and not for Scopus ($p = 0.39$). No significant differences were documented between Scopus and Web of Science regarding the number of total publications per year ($p = 0.07$). PubMed, as a service from the US National Library of Medicine (NLM), carries the highest number of publications. However, Web of Science is one of the most used search database, containing quality publications from editorials with high impact factor.

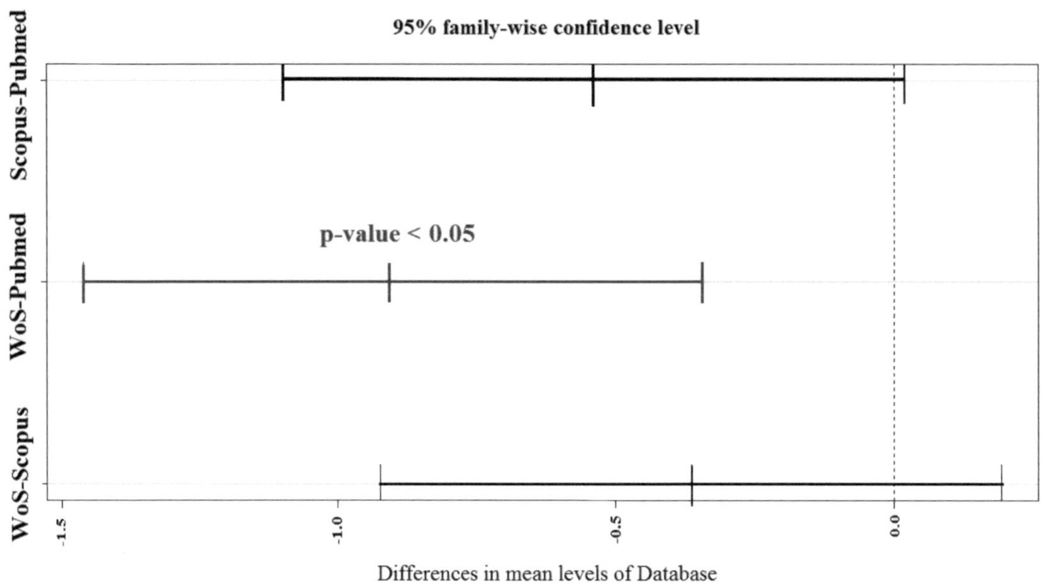

Figure 1. Plot based on Tukey's test showing the differences between the databases. There are significant differences between Web of Science and PubMed ($p = 0.0018$).

3.2. Study of the number of publications according to continents and countries

The location of the research related to the topic keywords according to continents and countries was also analyzed. Most of the countries in the world possess almost one publication along time. However, United States of America (USA) and China present the highest number of publications (111,392 and 89,849, respectively). The whole subset of countries with representativeness in the research about bioremediation using haloarchaea in soils, waters, and industrial raw is shown in **Figure 2**, with different intensities according to the abundance of publications. The language was not a representative deal since almost all the publications were written in English (97.4%). The diversity of the language in the different countries is remarkable despite of this, where other writing languages are German (0.5%), Spanish (0.4%), and Portuguese (0.4%).

As a further matter, all the data from the number of publications per country were grouped into continents for a more general view of the representativeness. The following results were obtained concerning total publications in the search: Asia (210,850), Europe (210,086), America (165,946), Oceania (19,038), Eurasia (18,752), and Africa (18,591). Africa has the lowest value of publications although the abundance of countries with positive results is large. Eurasia includes Turkey, Russia, Kazakhstan, and Georgia, which can be considered from Europe and Asia due to their geographical localization. Instead of including these countries on a single continent, a different class of continent is considered due to their high values of publications. The final values of Europe and Asia are so close with a difference of 800 publications, which implies a possible change of dominance in the next few years.

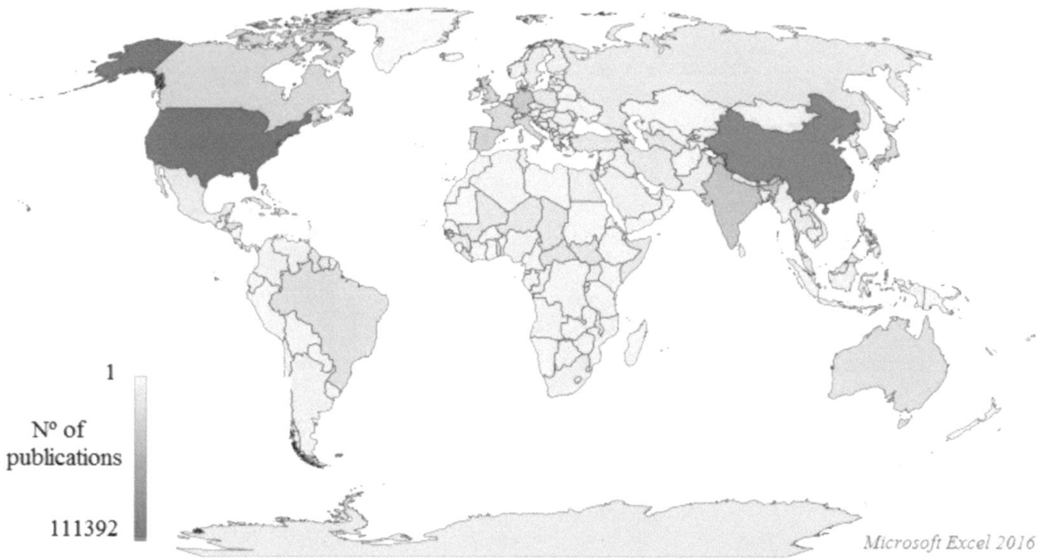

Figure 2. Global map representing the total publications per country by increasing intensity of color. Total number of publications is referred to the 10 keywords summation of results.

3.3. Impact of bioremediation using haloarchaea in literature

Haloarchaea were not directly related to bioremediation in literature until year 2007, and, even after this year, the number of publications has been low as can be noted in the percentage of occurrence of haloarchaea in bioremediation (0.06%). This value was calculated by analyzing the total publications per year regarding "bioremediation" as keyword in comparison to the combination of "bioremediation and haloarchaea or halophilic archaea."

The main lines of investigation regarding bioremediation using haloarchaea are developed by Bonete MJ (University of Alicante), Al-Mailem DM (University of Kuwait), and Martínez-Espinosa RM (University of Alicante). Furthermore, University of Alicante, University of Kuwait, and University of East Anglia are the most important institutions in terms of developing research lines about haloarchaeal species applied in bioremediation. These affiliations may be duplicated due to the fact some publications can contain two or more institutions in collaboration. Many of the authors with active experimental approaches in this way belong to the University of Alicante in collaboration with University of East Anglia (approximately, 50% of publications from the University of Alicante are in collaboration; 100% of publications from University of East Anglia are made in collaboration with University of Alicante).

Besides, the influence of the keywords "bioremediation" and "haloarchaea or halophilic archaea" was analyzed to delimit the total number of publications per year in the chosen databases. These results are plotted in **Figure 3**. There is an increasing tendency of the publications, with higher quantities of results during last years both in bioremediation (**Figure 3A**) and haloarchaea (**Figure 3B**). The keyword "halophilic archaea" shows a higher representativeness than "haloarchaea" in the search due to the presence of the word "halophilic" which can be

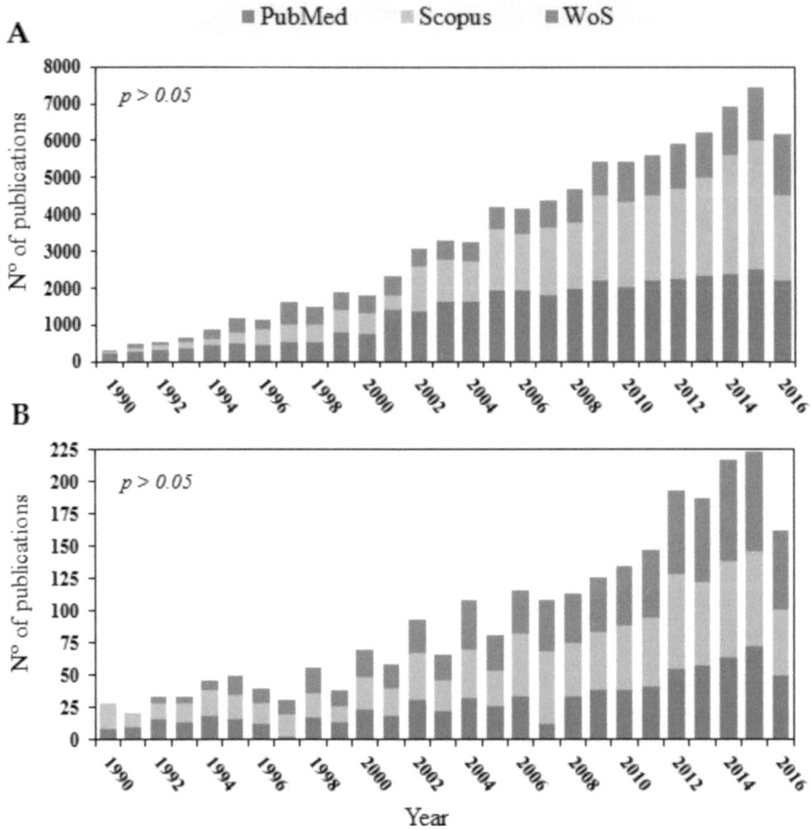

Figure 3. Graphical representation of the total number of publications per year from three databases (PubMed, Scopus, and Web of Science) concerning the keywords (A) "bioremediation"; and (B) "haloarchaea" and "halophilic archaea."

used for referring other species different from archaea (bacteria, microorganism, and plant; **Figure 3B**).

Both "bioremediation and haloarchaea" and "halophilic archaea" publications present a lower value of results (**Figure 4**). Since "haloarchaea" have a lower representativeness in literature than "bioremediation," the results show a limitation in the number of publications per year. Increasing results appear since year 2007, although some publications have been documented before in the way of potential studies using haloarchaea. Below year 2005, no guaranteed results concerning the topic are shown due to the lower interest in haloarchaea and the probable appearance of other microorganisms because of the use of "halophilic" in the search.

The impact of bioremediation research by countries is shown in **Figure 5**. This representation remarks the abundance of publications in the top 10 countries, where the USA is the dominant followed by China and India. This is not surprising owing to the presence of those countries in the top three of publications regarding all the keywords. In **Figure 6**, the top 10 countries that investigate in bioremediation using haloarchaea are shown, with Kuwait and Spain dominating the volume of total publications. Although Kuwait is not on the top 10 countries in

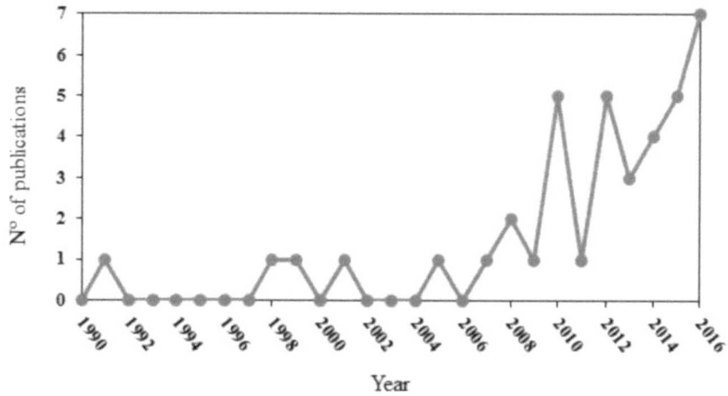

Figure 4. Graphical representation of the total publications per year PubMed, Scopus and Web of Science (WoS) databases using as keywords the average of "bioremediation and haloarchaea" and "bioremediation and halophilic archaea." Both combinations are represented as average due to their synonym relevance.

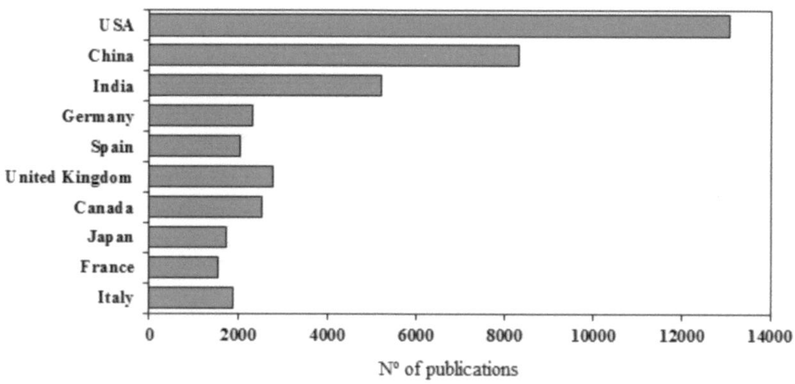

Figure 5. Main countries with the highest rate of publications in regard to bioremediation. Both Scopus and Web of Science results have been considered.

bioremediation publications, it is important to highlight its predominance in the study of haloarchaea as bioremediation tool.

Several areas of Kuwait are mainly characterized by hypersaline environments. Due to this fact, desalination is sometimes the only way of reliable water supply in most of the Arabian Gulf states. In South Kuwait, a desalination plant has showed a continued increase of salt, which can affect the local marine biodiversity. In 2011, it was suggested a distinct water quality approach to detect those issues without affecting the marine environment surrounding the water plant [38]. Desalination plants manage a huge quantity of salty water, although they imply several negative impacts. The main cause of desalination plant pollution in the environment includes noise pollution, visual pollution, reduction in recreational fishing and swimming areas, emission of materials into the atmosphere, and brine discharge [39]. Likewise, the climate is very like the one in Spain.

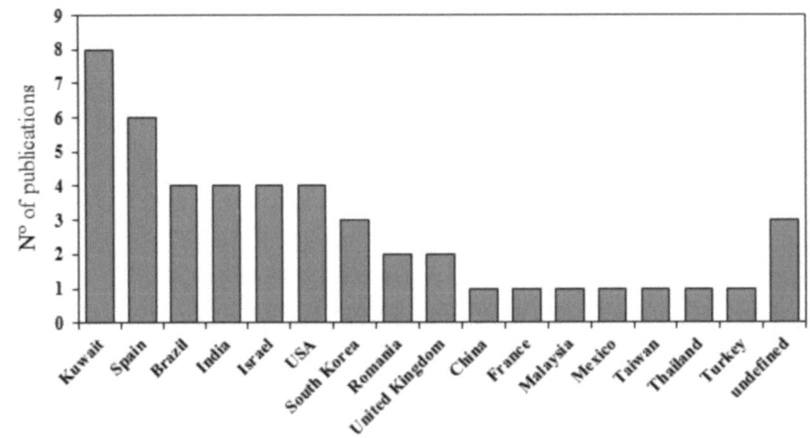

Figure 6. Mean total number of publications concerning the combination of keywords "bioremediation and haloarchaea" from Scopus and Web of Science search databases.

3.4. Bibliographic review of *Haloferax mediterranei*

Between years 2013 and 2016, there was an increase of interest for the study of this haloarchaeon and their possible applications. The highest number of publications concerning *Hfx. mediterranei* comes from Spain, with descriptions of different type of applications (industrial, environmental or medical) concerning this haloarchaeon. Authors like Bonete and Martínez-Espinosa (University of Alicante) or Pfeifer (Technische Universitat Darmstadt) appear as the first representative authors in this field, developing independent research lines in their institutions. In fact, the University of Alicante (Spain) is the first research institution in the ranking of number of publications about this haloarchaeon (**Figure 7**). Besides, those

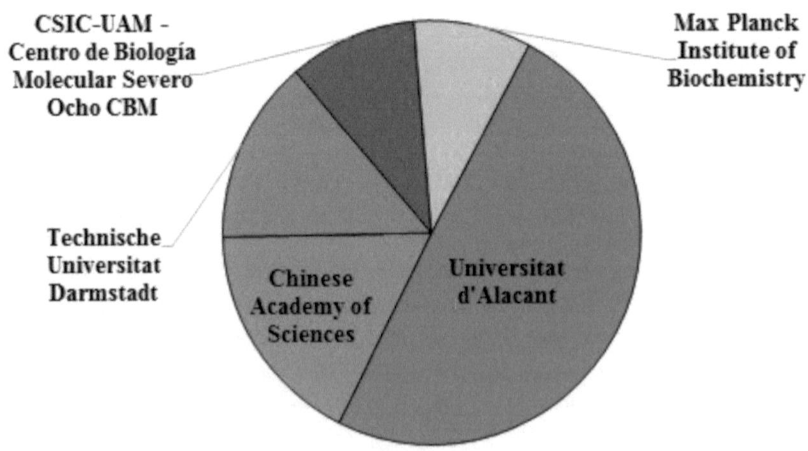

Figure 7. Top five of institutions/affiliations which are involved in research about *Hfx. mediterranei*: Universitat d'Alacant-University of Alicante (50%), Chinese Academy of Science (17%) Technische Universitat Darmstadt (14%), CSIC-UAM—Centro de Biología Molecular Severo Ochoa CBM (10%), Max Planck Institute of Biochemistry (9%). The publications may be duplicated in the way one article may be published by two affiliations.

authors are studying the biochemistry, molecular biology, and potential applications of *Hfx. mediterranei* since its discovery. This work is a contribution to that research group, the most involved in the understanding of *Hfx. mediterranei* in natural contaminated water and further applications [3, 25, 31].

4. Discussion

Globally, alternative remediation processes are gaining importance instead of chemical or solid methods in polluted compounds treatment. There is a global concern about wastewater and residue management due to the presence of toxic compounds [40]; thus, environmental friendly approaches are being financed by those countries with the higher rate of contamination, such as China or India [41]. Bioremediation, biostimulation, and bioaugmentation are suggested to be suitable alternatives and, due to this fact, they have received a great deal of attention in recent years [7].

Extremophiles may represent a good option for bioremediation of contaminated sites due to their biochemical and molecular properties. Halophilic microorganisms, especially haloarchaea, represent a highly promising way for removing polluted compounds thanks to their effective metabolism under extreme conditions or even in the presence of highly toxic compounds [2, 42]. Previous studies show the potential use of haloarchaea in bioremediation [3, 22]. Thus, a bibliographic review was developed along this work for the sole purpose to show the recent advances in bioremediation using these microorganisms. The recent gain of interest in potential haloarchaeal innovations is also shown. The source of information was based on three databases, some of which present high-quality impact in publications: PubMed, Scopus, and Web of Science. It was surprising the volume of publications from PubMed and Scopus in contrast to Web of Science, a scientific database containing high-quality publications and worldwide used. In fact, Web of Science is considered the most important database in terms of research evaluation of standards in some institutions such as universities, research institutes, or other institutions involved in research management.

Regarding to the topic, several countries are increasing their efforts in developing environmental friendly methods for removing contaminants; thus, the number of publications concerning bioremediation is highly increasing since recent years. The severe contamination documented in China and India may explain the high number of publications in benefit of bioremediation. The need to find solutions to this problem has increased the public and private finance for such projects. Bioremediation might be the solution for soils and water problems of pollution. The concerning about natural water contamination has always been present, but there is an increasing attention of creating soil quality remediation plans in China. Serious pollution risks were confirmed by the Chinese authorities in April 2014. About 16% of soils in China and 34.9% of brownfields exceed national standards for pollution with heavy metals and pesticides [43]. In India, in addition to the drinking water demand, there are evidences of groundwater contamination due to wastes of local industries. People experience symptoms of disease when drinking water from rivers, which reinforce the actual problem of water contamination [44].

Despite this, the diversity of metabolic pathways of *Hfx. mediterranei*, and in general haloarchaea, is greater than earlier assumed. The research about biochemical and molecular characteristics of haloarchaea will be a potential issue in biotechnological applications, especially on bioremediation of natural contaminated water. Furthermore, more investigation about the potential use of those microorganisms is needed to elucidate the optimum range of conditions for suitable bioremediation applications.

Author details

Sonia Aracil-Gisbert, Javier Torregrosa-Crespo and Rosa María Martínez-Espinosa*

*Address all correspondence to: rosa.martinez@ua.es

Departamento de Agroquímica y Bioquímica, Facultad de Ciencias, Universidad de Alicante, Alicante, Spain

References

[1] Breisha GZ, Winter J. Bio-removal of nitrogen from wastewaters–a review. Journal of American Science. 2010;**6**(12):508-528

[2] Le S, Paniagua D, Vazquez-Duhalt R. Biodegradation of organic pollutants by halophilic bacteria and archaea. Journal of Molecular Microbiology and Biotechnology. 2008;**15**:74-92. DOI: 10.1159/000121323

[3] Bonete M, Bautista V, Esclapez J, García-Bonete M, Pire C, Camacho M, Torregrosa-Crespo J, Martínez-Espinosa R. New uses of haloarchaeal species in bioremediation processes. In: Shiomi M, editor. Advances in Bioremediation of Wastewater and Polluted Soil. 1st ed. InTech; 2015. p. 23-49. DOI: 10.5772/60667

[4] Azubuike C, Chikere C, Okpokwasili G. Bioremediation techniques–classification based on site of application: principles, advantages, limitations and prospects. World Journal of Microbiology and Biotechnology. 2016;**32**:180. DOI: 10.1007/s11274-016-2137-x

[5] El S, Agathos S. Is bioaugmentation a feasible strategy for pollutant removal and site remediation? Current Opinion in Microbiology. 2005;**8**(3):268-275. DOI: 10.1016/j.mib.2005.04.011

[6] USA National Research Council. In Situ Bioremediation: When Does it Work? Washington, D.C.: National Academy Press; 1993. Available from: http://www.nap.edu/catalog/2131.html Accessed: March 8, 2017

[7] Adams G, Fufeyin P, Okoro S, Ehinomen I. Bioremediation, biostimulation and bioaugmentation: A review. International Journal of Environmental Bioremediation & Biodegradation. 2015;**3**(1):28-39. DOI: 10.12691/ijebb-3-1-5

[8] Cycoń M, Mrozik A, Piotrowska-Seget Z. Bioaugmentation as a strategy for the remediation of pesticide-polluted soil: A review. Chemosphere. 2017;**172**:52-71. DOI: 10.1016/j.chemosphere.2016.12.129

[9] Xu Y, Zhou N. Microbial remediation of aromatics-contaminated soil. Frontiers of Environmental Science & Engineering. 2016;**11**(2):1. DOI: 10.1007/s11783-017-0894-x

[10] Ivanov V, Hung Y. Applications of environmental biotechnology. In: Wang LK, Ivanov V, Hung Y, editors. Handbook of Environmental Engineering. Vol. 10: Environmental Biotechnology. CBS Publishers, New Delhi, India: Springer Science+Business Media; 2010. p. 1-27. DOI: 10.1007/978-1-60327-140-0_1

[11] Subhashini V, Swamy A. Phytoremediation of Pb and Ni contaminated soils using *Catharanthus roseus* (L.). Universal Journal of Environmental Research and Technology. 2013;**3**(4):465-472

[12] Zhang H, Tian Y, Wang L, Zhang L, Dai L. Ecophysiological characteristics and biogas production of cadmium-contaminated crops. Bioresource Technology. 2013;**146**:628-636. DOI: 10.1016/j.biortech.2013.07.148

[13] Jia X, Liu C, Song H, Ding M, Du J, Ma Q, Yuan Y. Design, analysis and application of synthetic microbial consortia. Synthetic and Systems Biotechnology. 2016;**1**(2):109-117. DOI: 10.1016/j.synbio.2016.02.001

[14] Nájera-Fernández C, Zafrilla B, Bonete M, Martínez-Espinosa R. Role of the denitrifying haloarchaea in the treatment of nitrite-brines. International Microbiology. 2012;**15**(3):111-119. DOI: 10.2436/20.1501.01.164

[15] Capes M, Coker J, Gessler R, Grinblat-Huse V, DasSarma S, Jacob C, Kim J, DasSarma P, DasSarma S. The information transfer system of halophilic archaea. Plasmid. 2011;**65**(2):77-101. DOI: 10.1016/j.plasmid.2010.11.005

[16] Arora S, Vanza M, Mehta R, Bhuva C, Patel P. Halophilic microbes for bio-remediation of salt affected soils. African Journal of Microbiology Research. 2014;**8**(33):3070-3078. DOI: 10.5897/AJMR2014.6960

[17] Cuadros-Orellana S, Pohlschröder M, Grossman M, Durrant L. Biodegradation of aromatic compounds by a halophilic archaeon isolated from the dead sea. Chemical Engineering Transactions. 2012;**27**:13-18. DOI: 10.3303/CET1227003

[18] Ventosa A, Nieto J, Oren A. Biology of moderately halophilic aerobic bacteria. Microbiology and Molecular Biology Reviews. 1998;**62**(2):504-544. 1092-2172/98/$04.0010

[19] Bonfá M, Grossman M, Mellado E, Durrant L. Biodegradation of aromatic hydrocarbons by haloarchaea and their use for the reduction of the chemical oxygen demand of hypersaline petroleum produced water. Chemosphere. 2011;**84**:1671-1676. DOI: 10.1016/j.chemosphere.2011.05.005

[20] Ghosal D, Ghosh S, Dutta T, Ahn Y. Current state of knowledge in microbial degradation of polycyclic aromatic hydrocarbons (PAHs): A review. Frontiers in Microbiology. 2016;**7**:1369. DOI: 10.3389/fmicb.2016.01369

[21] Oren A. Industrial and environmental applications of halophilic microorganisms. Environmental Technology. 2010;**31**(8–9):825-834. DOI: 10.1080/09593330903370026

[22] Bonete M, Martínez-Espinosa R. Enzymes from halophilic archaea: Open questions. In: Ventosa A, Orena A, Ma Y, editors. Halophiles and Hypersaline Environments. 1st ed. Berlin: Springer-Verlag; 2011. p. 359-373. DOI: 10.1007/978-3-642-20198-1_19

[23] Castillo Rodríguez F, Roldán Ruiz M, Blasco Plá R, Huertas Romera M, Caballero Domínguez F, Moreno-Vivián C, Martínez Luque-Romero M. Biotecnología Ambiental. 1st ed. Spain: Editorial Tébar; 2005. p. 377-387

[24] Oren A. Microbial life at high salt concentrations: phylogenetic and metabolic diversity. Saline Systems. 2008;**4**:2. DOI: 10.1186/1746-1448-4-2

[25] Torregrosa-Crespo J, Martínez-Espinosa R, Esclapez J, Bautista V, Pire C, Camacho M, Richardson D, Bonete M. Anaerobic metabolism in *Haloferax* genus: Denitrification as case of study. In: Poole RK, editor. Advances in Microbial Physiology. 1st ed. Vol. 68. Oxford Academic Press; 2016. p. 41-85. DOI: 10.1016/bs.ampbs.2016.02.001

[26] Valentine D. Adaptations to energy stress dictate the ecology and evolution of the Archaea. Nature Reviews Microbiology. 2007;**5**:316-323. DOI: 10.1038/nrmicro1619

[27] Munn C. Marine Microbiology: Ecology and Applications. 1st ed. Vol. 2004. Garland Science/BIOS Scientific Publishers; 2004. p. 81-120

[28] Gupta R, Naushad S, Baker S. Phylogenomic analyses and molecular signatures for the class *Halobacteria* and its two major clades: a proposal for division of the class *Halobacteria* into an emended order *Halobacteriales* and two new orders, *Haloferacales* ord. nov. and *Natrialbales* ord. nov., containing the novel families *Haloferacaceae* fam. nov. and *Natrialbaceae* fam. nov. International Journal of Systematic and Evolutionary Microbiology. 2015;**65**(3):1050-1069. DOI: 10.1099/ijs.0.070136-0

[29] Rodríguez-Valera F, Ruiz-Berraquero F, Ramos-Cormenzana A. Isolation of extremely halophilic bacteria able to grow in defined inorganic media with single carbon sources. Journal of General Microbiology. 1980;**119**(2):535-538. DOI: OO22-1287/80/0000-902

[30] Han J, Zhang F, Hou J, Liu X, Li M, Liu H, Cai L, Zhang B, Chen Y, Zhou J, Hu S, Xiang H. Complete genome sequence of the metabolically versatile halophilic archaeon *Haloferax mediterranei*, a poly(3-hydroxybutyrate-co-3-hydroxyvalerate) producer. Journal of Bacteriology. 2012;**194**(16):4463-4464. DOI: 10.1128/JB.00880-12

[31] Bonete M, Martínez-Espinosa R, Pire C, Zafrilla B, Richardson D. Nitrogen metabolism in haloarchaea. Saline Systems. 2008;**4**:9. DOI: 10.1186/1746-1448-4-9

[32] Lledó B, Martínez-Espinosa R, Marhuenda-Egea F, Bonete M. Respiratory nitrate reductase from haloarchaeon *Haloferax mediterranei*: Biochemical and genetic analysis. Biochimica et Biophysica Acta (BBA—General Subjects). 2004;**1674**(1):50-59. DOI: 10.1016/j.bbagen.2004.05.007

[33] Martínez-Espinosa R, Marhuenda-Egea F, Bonete M. Purification and characterization of a possible assimilatory nitrite reductase from the halophile archaeon *Haloferax mediterranei*. FEMS Microbiology Letters. 2001;**196**(2):113-118. DOI: 10.1111/j.1574-6968.2001.tb10550.x

[34] Martínez-Espinosa R, Richardson D, Butt J, Bonete M. Respiratory nitrate and nitrite pathway in the denitrifier haloarchaeon *Haloferax mediterranei*. Biochemical Society Transactions. 2006;**34**(1):115-117. DOI: 10.1042/BST0340115

[35] Martínez-Espinosa R, Lledó B, Marhuenda-Egea F, Díaz S, Bonete M. NO_3^-/NO_2^- assimilation in halophilic archaea: Physiological analysis, nasA and nasD expressions. Extremophiles. 2009;**13**(5):785-792. DOI: 10.1007/s00792-009-0266-y

[36] Martínez-Espinosa R, Richardson D, Bonete M. Characterisation of chlorate reduction in the haloarchaeon *Haloferax mediterranei*. Biochimica et Biophysica Acta (BBA)—General Subjects. 2015;**1850**(4):587-594. DOI: 10.1016/j.bbagen.2014.12.011

[37] Fang C, Ku K, Lee M, Su N. Influence of nutritive factors on C50 carotenoids production by *Haloferax mediterranei* ATCC 33500 with two-stage cultivation. Bioresource Technology. 2010;**101**(16):6487-6493. DOI: 10.1016/j.biortech.2010.03.044

[38] Uddin S, Al Ghadban A, Khabbaz A. Localized hyper saline waters in Arabian Gulf from desalination activity—an example from South Kuwait. Environmental Monitoring and Assessment. 2011;**181**(1–4):587-594. DOI: 10.1007/s10661-010-1853-1

[39] Tularam G, Ilahee M. Environmental concerns of desalinating seawater using reverse osmosis. Journal of Environmental Monitoring. 2007;**9**(8):805-813. DOI: 10.1039/b708455m

[40] Toxics Action Center. The Problems with Waste. [Internet]. 2015. Available from: http://www.toxicsaction.org/problems-and-solutions/waste [Accessed: May 21, 2017]

[41] Numbeo. Pollution Index for Country 2017. [Internet]. 2017. Available from: https://www.numbeo.com/pollution/rankings_by_country.jsp [Accessed: May 21, 2017]

[42] Falb M, Müller K, Königsmaier L, Oberwinkler T, Horn P, von S, González O, Pfeiffer F, Bornberg-Bauer E, Oesterhelt D. Metabolism of halophilic archaea. Extremophiles. 2008;**12**(2):177-196. DOI: 10.1007/s00792-008-0138-x

[43] Yang H. China's soil plan needs strong support. Nature. 2016;**536**:375. DOI: 10.1038/536375a

[44] Harris T. Rains or not, India faces drinking water crisis. Phys.org. [Internet]. 2016. Available from: https://phys.org/news/2016-06-india-crisis.html [Accessed: May 12, 2017]

Chapter 5

Laboratory-Scale Biodegradation of Fuel Oil No. 6 in Contaminated Soils by Autochthonous Bacteria

Hilda Amelia Piñón-Castillo,
Daniel Lardizabal Gutiérrez,
Francisco Javier Zavala-Díaz de la Serna,
Daniel Hernández-Castillo,
Laila N. Muñoz-Castellanos,
Blanca E. Rivera-Chavira and
Guadalupe Virginia Nevárez-Moorillón

Additional information is available at the end of the chapter

http://dx.doi.org/10.5772/intechopen.70350

Abstract

In order to evaluate the degradation of fuel oil no. 6 (FO6) in contaminated soil, laboratory-scale bioreactors were set up to study biostimulation, bioaugmentation, and natural attenuation processes. A solution of fertilizers was added in biostimulation and bioaugmentation (0.03% N, 0.01% P). To the bioaugmentation process, an enrichment culture of indigenous hydrocarbon-degrading microorganisms was also added once a week. Total aerobic and hydrocarbon-degrading microorganisms were determined by plate count, and total petroleum hydrocarbon (TPH) concentration was determined gravimetrically (EPA method 9071b) every 15 days. After 1 year of study, degradation rate was higher for biostimulation (0.19 g TPH/day), followed by natural attenuation (0.18 g TPH/day) and bioaugmentation (0.16 g TPH/day). TPH showed a change in composition of hydrocarbons, attributed to microbiological activity. Microbial counts of hydrocarbon-degrading microorganisms were on the range of 4–6 log CFU/g soil. Preliminary bacterial identification corresponded to *Pseudomonas*, *Rhodococcus*, *Actinomyces*, and *Bacillus* strains; randomly amplified polymorphic DNA (RAPD); and denaturing gradient gel electrophoresis (DGGE) analysis demonstrated a large microbial diversity. From the degradation rates, it can be predicted that such limits will be achieved by increasing further 107–117 days of the treatments. Results demonstrated to be efficient on the restoration of contaminated soil, being an alternative to treat soils contaminated with heavy hydrocarbons.

Keywords: fuel oil no. 6, biostimulation, bioaugmentation, natural attenuation, PCR-DGGE

1. Introduction

The Food and Agricultural Organization (FAO) defines soil degradation as the loss of soil productivity characteristics, either quantitatively or qualitatively. This process can be consequences of hydraulic progression or wind erosion, nutrient depletion, desertification, salinization, and pollution, among other factors [1]. Soil conservation is necessary since many important life-related processes take place in this ecosystem, including food production, biogeochemical cycles, industrial uses, and mineral extraction, besides being the substrate of our living environment. Regrettably, these human activities are the main source of soil pollution [2–5].

Exploitation and production of petroleum products, and our dependency on petroleum hydrocarbon exploitation, have brought an increase in pollution problems due to their recalcitrant nature [6–8]. The release of petroleum hydrocarbons into environment, either accidentally or as result of human-related activities, is one of the most deleterious causes of soil and water contamination, since they can cause a severe damage to the environment. This damage includes problems related to plant development and toxicity for microbial and animal species. Additionally, toxic effects could be derived from lixiviation of petroleum products into groundwater, or bioaccumulation, which can introduce the contaminant into the food web and eventually be eaten by humans [9–12].

During the last decades, several soil restoration techniques have been proposed, including extraction of contaminants by mechanical methods, by solvent extraction, or by using the capacity of soil microorganisms to degrade hydrocarbons (bioremediation) [2, 13, 14]. Physicochemical restoration techniques can be done in situ, or the contaminated soil can be removed and the treatment can be done ex situ. Among physicochemical restoration techniques, mechanical separation, electro remediation, cofferdam system, and soil cover/insulation can be included [10, 11, 15]. The objective of chemical remediation is to convert the contaminant into a substance with lower toxicity; soil washing, immobilization, chemical and photochemical reduction, and soil flushing are techniques included in this type of treatment. Both physical and chemical techniques have high associated costs and cannot remove the contaminant completely from the site [9–12].

On the other hand, bioremediation has proven to be a cost-effective, environmental-friendly, and simple technique to restore hydrocarbon-contaminated soil sites [16, 17]. Bioremediation techniques include the use of nutrients to increase microbial activity (biostimulation), the addition of an enrichment culture of contaminant-degrading microorganisms (bioaugmentation) or monitoring of the place, sometimes with air injection to promote the natural degradation process (natural attenuation). Limitations encountered in bioremediation include nutrient availability, but they can be supplied as inorganic salts [2, 5]. For hydrocarbon biodegradation in soil, the use of earthworms [16], rhizospheric microorganisms [18], the addition of surfactants and fertilizers [19], composting [20, 21], and landfarming as well as the effect of the type of soil or temperature have been evaluated [20, 22].

In México, it has been reported that as a result of years of activity of the basic chemical industry, petrochemicals, mining, and hydrocarbon refining, an area of 25,967 km^2 of degraded soil was reported in 1999 [6]. In the northern state of Chihuahua, Mexico, an accidental release of petroleum hydrocarbons was reported in 2010 as consequence of a clandestine poaching of a hydrocarbon pipeline [8]. In one of the companies that is settled in Chihuahua City, a spill of fuel oil no. 6 from an underground storage tank, which was leaking from non-hermetic joints, was discovered during a volunteer environmental auditing process by PROFEPA (Mexican Federal Environmental Authority). Site restoration was requested, to achieve a maximum of 3.5 g/kg total petroleum hydrocarbons (TPH), in agreement with the Official Mexican Standard that refers to the levels of soil contamination allowed in industrial settings and sites that are contaminated (NOM-EM-ECOL-138-2002) [23].

Fuel oil no. 6 (FO6) is one of the last products obtained in petroleum distillation and is reported as saturate, aromatic, resin, and asphaltene (SARA) fraction, with a concentration of around 25% of saturated hydrocarbons, 35% of aromatics hydrocarbons, 20% of resins, and 20% of asphaltenes [24, 25].

The aim of this work was to evaluate at laboratory scale, and three biodegradation treatments for soil heavily contaminated with fuel oil no. 6 were done: (1) natural attenuation, (2) Biostimulation, and (3) bioaugmentation; also, the influence of the different treatments on the soil microbiota was evaluated.

2. Materials and methods

2.1. Microcosm construction

Three laboratory-scale bioreactors were constructed, to hold 6.5 kg of soil contaminated with fuel oil no. 6. An air distribution system with a flexible pipeline was built and incorporated into each microcosm. The microcosm set up as natural attenuation (control) was irrigated with 100 mL of water every 48 h. The biostimulation microcosm was irrigated every 48 h with 100 mL of a fertilizer solution containing ammonium phosphate and urea to a final concentration of 0.03% (w/v) nitrogen and 0.01% (w/v) phosphorous. In order to evaluate bioaugmentation system, a 100 mL enriched culture of soil microorganisms grown in a system with FO6 as carbon source and the same fertilizer solution were added once a week. The enriched culture was prepared adding 5 g of FO6 contaminate soil in 200 mL of fertilizer solution; FO6 saturated air was bubbled into the system (**Figure 1**).

2.2. Physicochemical soil characterization

Soil contaminated with FO6 was characterized by standard techniques, to determinate pH, texture, apparent density, nitrate, and phosphate, among other characteristics. Heavy metals were quantified by flame atomic absorption (Model Avantel Sigma, GBC, Hampshire, IL, USA); As, Hg, and Se were determined by FAAS-HG, using NIST certified standards.

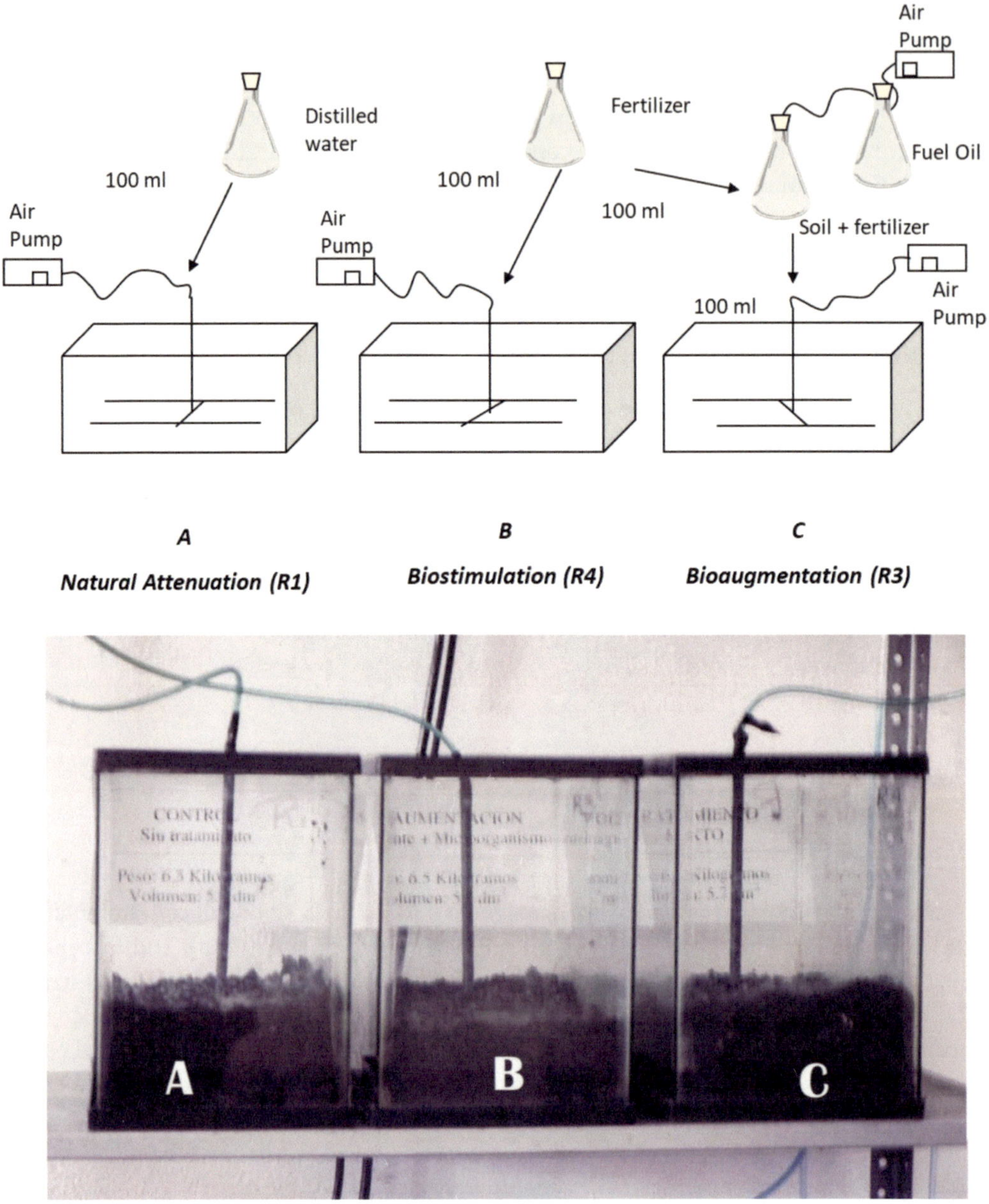

Figure 1. Lab-scale bioreactor construction. (A) Natural attenuation, (B) biostimulation, and (C) bioaugmentation.

2.3. Microbial quantification of total mesophilic bacterial count and hydrocarbon-degrading mesophilic bacteria

The quantification of total mesophilic aerobic microorganisms, as well as hydrocarbon-degrading aerobic microorganisms, was done every 15 days for 1 year. A composite soil sample (20 g) was obtained from 2 to 5 cm depth into the soil column at each bioreactor. Standard plate-counting method using serial dilution and spread plate technique were used. Standard Methods Agar (Bioxon, Mexico) was used for total mesophilic count, and for hydrocarbon-degrading bacteria, M9 minimal agar without carbon source added was utilized [23]. Plates were incubated in a closed chamber with a FO6-saturated atmosphere at 28°C for 48 h in the case of total mesophilic count and 7–10 days for hydrocarbon-degrading microbial counts. Results were expressed as log CFU/g of soil. From the hydrocarbon-degrading counts, representative colonies were isolated and streaked into nutrient agar (Bioxon, Mexico) to obtain isolated colonies. Macroscopic and microscopic morphology, Gram stain, as well as oxidase and catalase biochemical tests were employed for a preliminary identification of the isolates. Isolated strains were evaluated for FO6 degradation capacity using an M9 agar without carbon source added and with a top layer of FO6; bacteria were directly inoculated into the agar using a sterile toothpick, and growth was determined after 5 days of incubation at 28°C [26].

2.4. Randomly amplified polymorphic DNA (RAPD) analysis and denaturing gradient gel electrophoresis (DGGE) of hydrocarbon-degrading bacterial isolates

Isolated bacterial strains of hydrocarbon-degrading bacteria were grown in trypticase soy agar (Bioxon, Mexico), and after an incubation of 12 h at 28°C, cells were collected by centrifugation; 200 μL of lysis buffer (Tritón X-100 at 2%, SDS 1%, NaCl 10 mM, EDTA 1 mM) and 200 μL of PBS-phenol-saturated solution were added. The samples were mixed in vortex for 4 min and incubated at 65°C for 1 h. After incubation, 200 μL of TE was added, and tubes were centrifuged; the aqueous phase was transferred to a clean tube. DNA in the aqueous phase was precipitated with ethanol 70% with incubation at −20°C overnight and was later recovered by centrifugation and suspended in 400 μL of TE added with 30 μg of aRNAs. Samples were incubated at 37°C for 5 min, and the DNA was precipitated with isopropanol and recuperated by centrifugation. The pellet was air-dried and suspended in molecular biology water. This DNA was used for PCR amplification and DGGE and RAPD analysis [27]. PCR amplification for DGGE analysis was performed using primers MAR (5′ GCG CCG CCG CGC GCG GCG GGC GGG GCG GGG GCA CGG GGG GCC TAC GGG AGG CAG CAG 3′) and MAR 1 (5′ ATT ACC GCG GCT GCT GG 3′). Amplification mixture was prepared with buffer PCR (1×), $MgCl_2$ (3 mM), dNTPs (0.2 mM), primers (0.4 pM), Taq polymerase (1 U/μL), and DNA (100 ng). Reactions were processed with an initial denaturation step (92°C for 5 min) followed by 35 cycles of denaturation (92°C for 45 s), annealing (55°C for 30 s), and elongation (72°C for 45 s) and then a final elongation step (72°C for 5 min).

DGGE analysis was done using 1-mm-thick, 16 × 16 cm polyacrylamide gels [7.0% (v/v) acrylamide-bisacrylamide (37.5/1); denaturant (urea/formamide)]. From the PCR products, 25 μL was added on denaturing gradients of 58–50%. A mixture of 7 M urea and 40% formamide was defined as 100% denaturant. The gels were run in 0.5× TAE (40 mM Tris-acetate and 1 mM EDTA) at 45 V for 20 h at 60°C. Thereafter, the gel was stained with silver nitrate (silver satin plus Bio-Rad).

For the RAPD analysis, primers F (5′-GGACTGCAGA-3′) and R (5′-AGCTGACCGT-3′) were used. Amplification mixture included buffer PCR (1×), $MgCl_2$ (3 mM), dNTPs (0.2 mM), primers (0.4 pM), Taq polymerase (1 U/μL), and DNA (100 ng). PCR was processed with 42 cycles of denaturation (92°C for 1 min), annealing (36°C for 1.5 min), and elongation (72°C for 2.5 min) and then a final elongation step (72°C for 5 min). PCR products were analyzed in agarose gel (1.5%) in TBE (45 mM Tris-borate/1 mM EDTA). The gel was stained with ethidium bromide solution (0.5 μg ml^{-1}).

2.5. Hydrocarbons quantification

Total petroleum hydrocarbon (TPH) concentrations were determined by the 8440 EPA method gravimetric method [28], as well as by the 3570 EPA method, based in Fourier transform infrared (FT-IR) analysis [29]. For the gravimetric method, to 5 g of Na_2SO_4-dried soil sample, 25 mL of methylene chloride was added, and the mixture was shaken at 35 RPM for 1 h. Later, samples were filtered in fiberglass; the dissolvent was evaporated, and the sample was weighted. Extracted hydrocarbons obtained from the previous step were used for FT-IR quantification, by adding 4 mL of tetrachloroethylene, and evaluated in the total hydrocarbon analyzer (Model HC-404, Buck Scientific, East Norwalk, CT, USA). A standard curve was prepared using 5000–20,000 mg/kg of n-hexadecane as standard; lecture was taken at 2927.37 cm^{-1} peak. Also, a qualitative FT-IR analysis was done to determine the presence of functional groups that can describe the compounds present in FO6 (Model Spectrum GX, Perkin Elmer, Waltham, MA, USA).

2.6. Hydrocarbon degradation analysis by differential thermal analysis (DTA) and thermal gravimetric analysis (TGA)

In order to analyze the change in hydrocarbon composition, DTA and TGA analyses were done from samples obtained in the gravimetric TPH analysis. Samples were analyzed from room temperature to 700°C at a rate of 15°C/min in an air atmosphere in a TGA instrument (Model 2960, TA Instruments, New Castle, DE, USA).

3. Results and discussions

3.1. Physicochemical soil characterization

The soil profile of the FO6-contaminated soil was of an alkaline (pH 8.9) loam soil (sand 38%, silt 29%, clay 32%). Heavy metal concentrations were low (**Table 1**). Selection of remediation strategies depends largely on the extrinsic characteristic of the contaminated site, such as pH, temperature, and nutrient concentration, among other factors [3, 30], since the hydrocarbon biodegradation rate is important. The pH value depends on the organic material content and soil nature; a basic pH in soil contaminated with hydrocarbons has been found. Soil characteristics such as hydraulic conductivity, water retention, and porosity are also important for nutrient and air distribution that are essential for microbial activity. According to the results presented in **Table 1**, hydraulic conductivity is very low. It has been described that soil with high hydrocarbon concentrations presents changes in water retention and hydraulic conductivity properties [6, 31–33]. This phenomenon interferes with biological nutrient cycling, because biological remediation process depends on nutrient distribution for microbial growth.

Physicochemical parameters	pH ($CaCl_2$ 0.01 M)	8.98
	Hydraulic conductivity (cm/h)	0.56
	Electric conductivity (mmhos/cm)	10
Inorganic ion content	N–NO_3 (kg/ha)	367
	P (kg/ha)	12.52
	$CaCO_3$ (%)	24.94
	K (ppm)	53.75
Textural soil classification	Clay (%)	32
	Silt (%)	29
	Sand (%)	38
		Clayey
Metal content	As (ppm)	0.01
	Ba (ppm)	3.6
	Cd (ppm)	<0.025
	Ni (ppm)	<0.1
	Hg (ppm)	<0.005
	Ag (ppm)	<0.1
	Pb (ppm)	<0.1
	Se (ppm)	<0.001

Table 1. Physicochemical characterization of soil contaminated with fuel oil no. 6.

3.2. Microbial growth in the lab-scale bioreactors and preliminary identification of hydrocarbon-degrading microorganisms

Microbial counts of total mesophilic and hydrocarbon-degrading microorganisms were evaluated in the three different lab-scale bioreactors every 15 days. The number of microorganisms was similar in the three systems, as observed in **Figure 2**, but in all treatments, an increase in microbial numbers was observed. It was also evident that a large proportion of the microbial population was able to grow using petroleum hydrocarbons as carbon source, as observed by the number of hydrocarbon-degrading bacteria (**Figure 2**). All treatments showed an increase in the microbial count up to the day 280 of treatments, and a slight decrease was observed during the rest of the study.

In order to characterize the diversity and similarity of hydrocarbon-degrading microorganisms present in the different treatments, bacterial strains with differences in morphological characteristics were isolated (Gram stain), and their distribution along the experiment is described in **Figure 3**. The higher microbial diversity was found in the first and the last trimester. The initial metabolic characterization (oxidase and catalase tests) along with morphological characterization showed the presence of *Bacillus*, *Pseudomonas*, *Rhodococcus*, and *Actinomyces* strains.

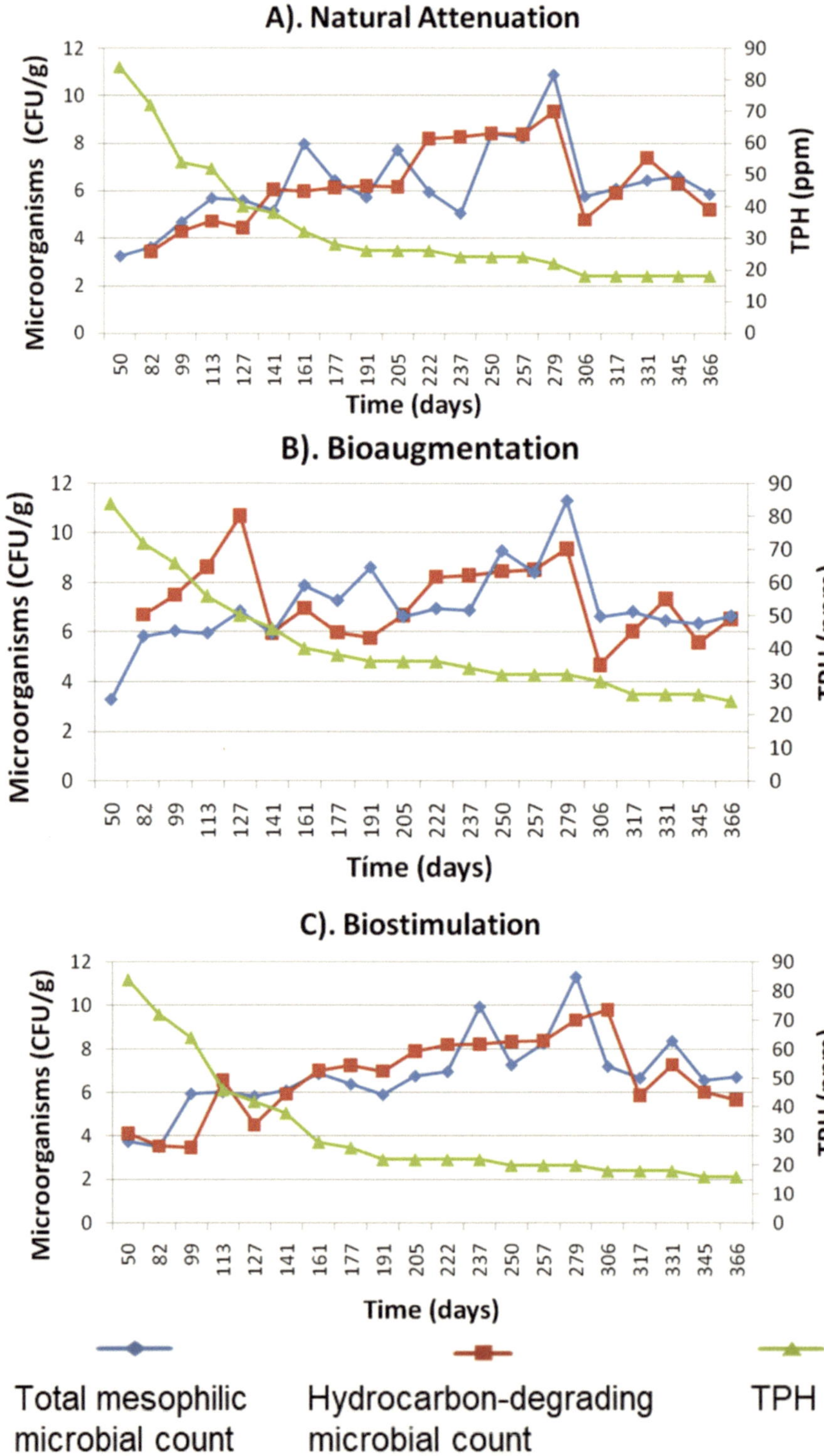

Figure 2. Microbial growth (total mesopilic and hydrocarbon-degraders) and hydrocarbon degradation (TPH) in lab-scale bioreactors. (A) Natural attenuation, (B) bioaugmentation, and (C) biostimulation.

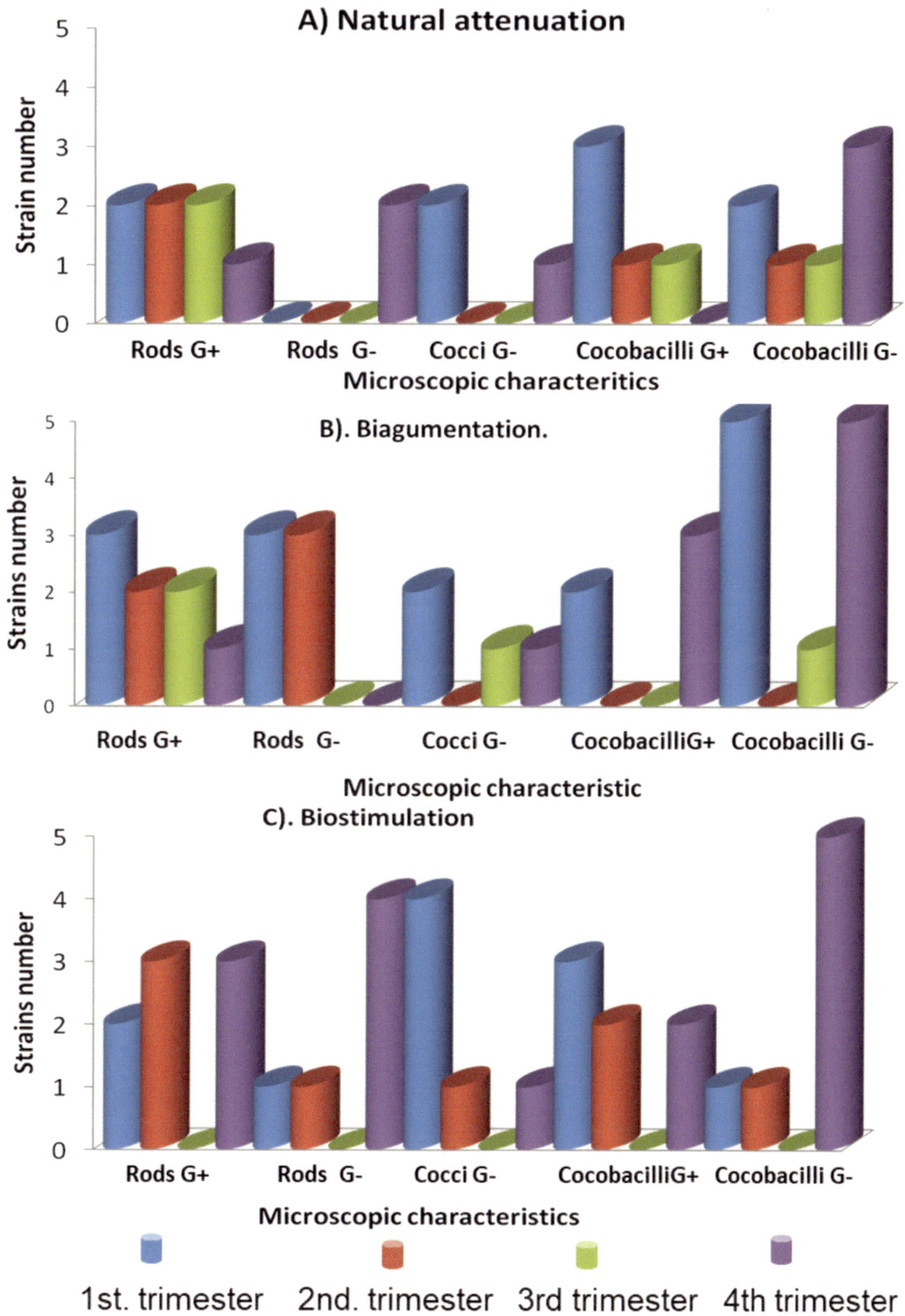

Figure 3. Preliminary characterization of microbial strains isolated from lab-scale bioreactors. (A) Natural attenuation, (B) bioaugmentation, and (C) biostimulation.

For DGGE and RAPD analysis, bacterial strains representative of the different treatments and at different sampling times were selected, to verify the diversity of microorganisms present. Results are shown in **Figure 4**, where strain identification shows the bioreactor (R1, natural attenuation; R3, bioaugmentation; R4, biostimulation), as well as the sampling period (M). As shown in **Figure 4**, the RAPD analysis and DGGE analysis present a high diversity in all the strains isolated from the different treatments. The RAPD analysis showed that strains in lines 3 and 5 have the same pattern, but the rest of the strains showed a different pattern, indicating that microbial strains are different (**Figure 4A**). However, DGGE analysis showed a high similarity as shown in the lines 5–7 and 15–18 (**Figure 4B**). The differences are due to differences on the DNA regions amplified in each technique.

It has been reported that microorganisms from the *Pseudomonas* group had a big pathway of degradation for many types of pollutants [34]. This microbial group can degrade a wide variety of hydrophilic and hydrophobic compounds, including polycyclic aromatic hydrocarbons

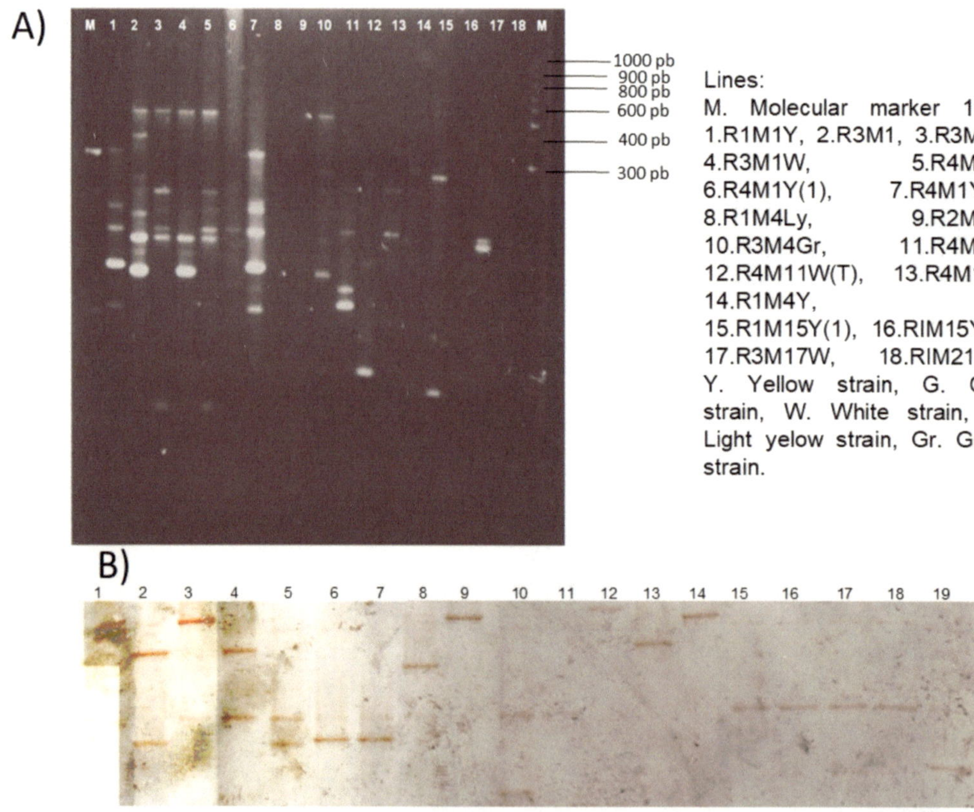

Lines:
M. Molecular marker 1kbp, 1.R1M1Y, 2.R3M1, 3.R3M1G, 4.R3M1W, 5.R4M1W, 6.R4M1Y(1), 7.R4M1Y(2), 8.R1M4Ly, 9.R2M4W, 10.R3M4Gr, 11.R4M4W, 12.R4M11W(T), 13.R4M11Y, 14.R1M4Y, 15.R1M15Y(1), 16.RIM15Y(2), 17.R3M17W, 18.RIM21Ly2. Y. Yellow strain, G. Gray strain, W. White strain, Ly. Light yelow strain, Gr. Green strain.

Lines: 1Lines: 1.R3M4Gr, 2.R4M4W, 3.R4M11W(T), 4.R4M11Y, 5.R1M4Y, 6.R1M15Y(1), 7.R1M15Y(2), 8.R3M17W, 9.R4M15W, 10.R4M15O, 11.R4M17R, 12.R1M21Ly2, 13.RIM21Ly3, 14.R3M21Ly1, 15.R4M21Ly4, 16.R1M1O, 17.R3M4O, 18.R1M11R, 19.R4M21Ly1. Y. Yellow strain, G. Gray strain, W. White strain, Ly. Light yelow strain, Gr. Green strain. O. organe strin , R.red strain

Figure 4. Analysis of hydrocarbon-degrading bacterial strains isolated from the different biodegradation treatments (A) DGGE and (B) RAPD. Strain identification is related to the bioreactor (R1, natural attenuation; R3, bioaugmentation; R4, biostimulation) as well as the sampling period (M); they are also identified by the color of the colony.

(PAHs) [35–37]. On the other hand, *Bacillus*, *Rhodococcus*, and *Actinomyces* strains can also degrade PHAs [37–39]. The isolated strains from the soils of the three types of experiments were cultivated in M9 minimum agar added with FO6, in order to observe if they can use TPH present in FO6 as carbon source (**Figure 5**). TPH utilization was visible by a zone of degradation around the bacterial growth; in some cases, emulsification of TPH was observed, due to the production of biosurfactants by the microorganism [19, 34]. It has been reported that in hydrocarbon-degrading microorganisms, biosurfactant production allows strains to make the hydrophobic carbon source available for biodegradation [19, 40–42]. This is especially important in the degradation of complex recalcitrant compounds, such as the ones present in FO6.

3.3. Total petroleum hydrocarbon degradation

In order to evaluate the efficiency of each treatment, TPH was quantified along the time of biological treatment. Two methods were compared for TPH quantification, the gravimetric method and the FT-IR method; results showed a correlation between both methods, with a highly significant correlation coefficient ($r = 0.97$, $p < 0.01$). In the three biological treatments, TPH diminished with respect to time of treatment (**Figure 2**). The higher percentage of TPH degradation was for bioaugmentation (86.6%), followed by biostimulation (86.6%) and finally by natural attenuation (85.5%). For the three treatments, a two-rate of TPH degradation process was observed, with a higher degradation rate during the first 145 days and a slower rate during the rest of the experiment (**Table 2**). TPH values achieved in the year of study were

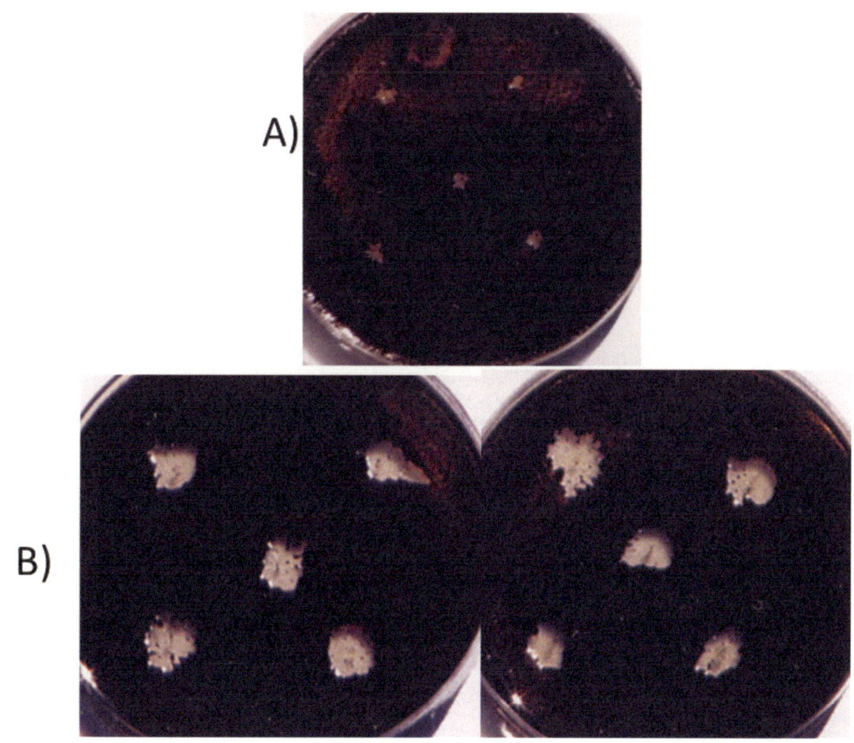

Figure 5. Analysis of hydrocarbon-degrading bacterial strains isolated from the different biodegradation treatments A) Hydrocarbon degradation but no biosurfactant production. B) Hydrocarbon degradation and biosurfactant production.

near to the values allowed by Mexican regulations [43] for soil. The rate of degradation was better in the first 145 days in all the microcosms, and in the last period of time, this value decreases (**Table 2**). These results showed that in the first 145 days of process the better treatment is bioaugmentation; nevertheless, their rate decreased in the last days.

It has been described that the use of different bioremediation strategies such as bioaugmentation, natural attenuation, or biostimulation does not produce differences in the decrease of TPH in contaminated sites [17]. On the contrary, it has been reported that bioaugmentation was the second best treatment for the removal of hydrocarbons in contaminated sites and that natural attenuation was the less effective [7]. The use of plants in combination with other biodegradation techniques, such as bioaugmentation, landfarming, assisted phytoremediation, or phytoremediation, has shown to have TPH removal efficiencies of 86, 68, 68, and 59%, respectively [7, 22, 44]. In addition, the use of compost or food waste to increase nutrient availability has also showed to be efficient in TPH removal from soil [45].

The adsorption capacity of soil, to integrate organic compounds into their particles, has been described. This function restricts the availability of contaminants to microorganisms for degradation and is, therefore, a conditional step for biodegradation: the pollutant cannot be degraded faster than it is liberated from the soil particles [46, 47]. Also, degradation can be related to microbial growth. In **Figure 2**, microbial growth of hydrocarbon-degrading bacteria decreased, and at the same time, the rate of hydrocarbons degradation was slower. The relation directly between TPH and microbial growth has been reported [48].

3.4. Hydrocarbon characterization

In order to evaluate the change in TPH composition throughout the experiment, a TGA and DTA analysis was done. The analysis of TPH found in soil at the beginning of the experiment showed a decrease in weight at 500°C and remains constant until 550°C; another weight decrease is evident until 625°C. DTA analysis showed a series of bands in the range of 450–500°C (**Figure 6A**). These results indicated that TPH was composed of hydrocarbons with aliphatic chains; this result is similar with the description done in the Prestige oil spill, where the presence of PAH was described in FO6 [25].

	Rate of biodegradation of HTP (mg of HTP/day)			Time (days) required for closure*
	0–145 days of experimentation	145–366 days of experimentation	0–366 days of experimentation	
Natural attenuation	207.6	50.108	112.5	96
Bioaugmentation	225.3	42.5	114.9	93
Biostimulation	208.6	51.21	113.5	86

*Time (days) required to reach the value of HTP specified by Mexican normativity [43].

Table 2. TPH degradation in a soil contaminated with fuel oil no. 6, subjected to three bioremediation treatments at lab-scale bioreactors.

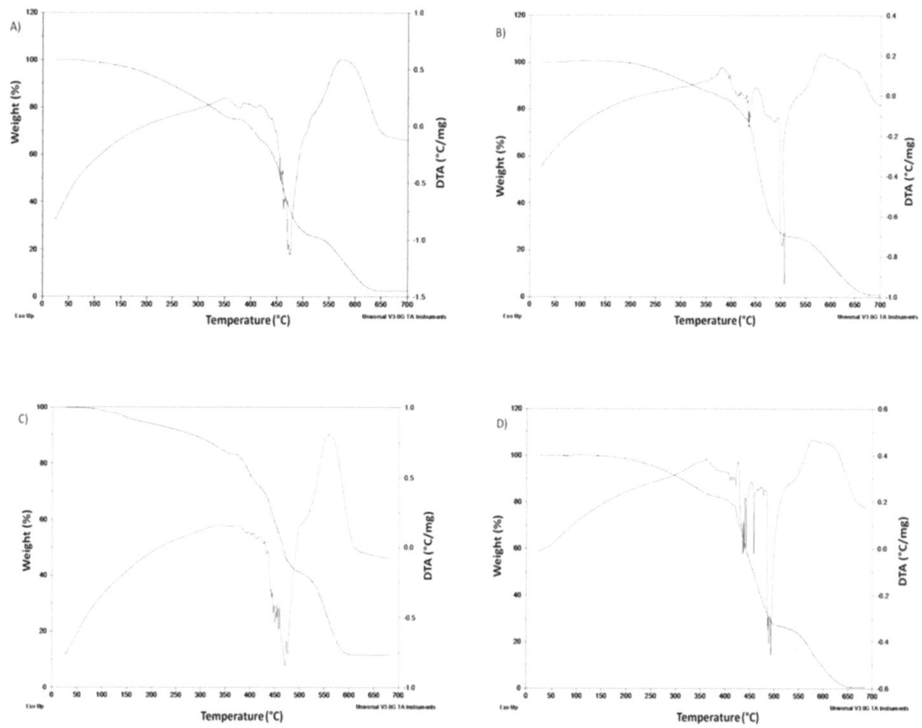

Figure 6. Analysis of TPH by DTA and TGA in the different treatments. (A) Initial TPH analysis. TPH remaining after 366 days of biological treatment, (B) natural attenuation, (C) bioaugmentation, and (D) biostimulation.

The results of the DTA and TGA analysis of the remaining TPH after 366 days of biological treatment indicated differences among bioremediation techniques used. Natural attenuation showed bands in the range of 425–475°C, with a bigger number of signals than at the beginning of the experiment (**Figure 6B**). In the case of bioaugmentation and biostimulation (**Figure 6C** and **D**, respectively), results were similar. The increase in the number of bands and the shift in degradation temperatures are the result of transformation of large hydrocarbon molecules into others of lower molecular weight. On the other hand, once the shorter molecules are consumed in the contaminated site, the use of more complex molecules is initiated; this is reflected in the TGA and DTA analysis by the decrease in evaporation temperature in the bands.

On the other hand, FT-IR analysis not only showed TPH decrease in concentration, but was also used to establish a relationship between hydrocarbons with CH_3—CH_2 groups and the presence of CH_3 residues. The band at 1459.52 cm^{-1} is the representative of the CH_3—CH_2 link, and the band in 1377.03 cm^{-1} is the representative of CH_3 residue. The relation between the CH_3 and CH_2/CH_3 can be used to express the conversion of compounds with differences in the proportion of those functional groups. Results are shown in **Table 3**, where an increase of CH_3 is observed in all treatments.

Treatment	Time of treatment (days)	CH_2–CH_3/CH_3 ratio
Natural attenuation	0	1.70
	93	1.19
	184	1.12
	275	1.13
	366	1.14
Bioaugmentation	0	1.70
	93	1.18
	275	1.10
	366	1.15
Biostimulation	0	1.70
	93	1.19
	184	1.14
	366	1.20

Table 3. FT-IR analysis of TPH from the three lab-scale bioreactors for bioremediation strategies, in soil contaminated with FO6. The relationship in band height (1459.52 cm^{-1}/1377.03 cm^{-1}) is shown as CH_2–CH_3/CH_3 ratio.

Relationship of functional groups has been used for the determination of the relation between C and H [42]. The change in bands in the TGA and DTA analyses, as well as the increase in CH_3 observed by FT-IR, suggests that the transformation of TPH was mediated by the soil microbiota. Bioremediation strategies such as the ones used in this study have been successfully used in site restoration [48–50].

4. Conclusion

Soil is a nonrenewable resource; it is therefore important to generate restoration strategies in sites that have been contaminated due to human activities, where soil quality and functionality need to be restored. Bioremediation of soil contaminated with hydrocarbons can be a time-consuming technique but is cost effective and environmentally appropriate, since the contaminants are transformed by soil microorganisms into simple noncontaminant organic compounds. In this work the degradation of TPH was evaluated by three treatments: natural attenuation, biostimulation, and bioaugmentation. After 1 year of study, degradation was higher for biostimulation (0.19 g TPH/day), followed by natural attenuation (0.18 g TPH/day) and bioaugmentation (0.16 g TPH/day). *Pseudomonas, Rhodococcus, Actinomyces,* and *Bacillus* were present in the reactors, according to preliminary identification of bacteria, with a large microbial diversity, as observed by molecular culture-independent techniques. Based on qualitative analysis of TPH, a change in the length of the aliphatic chain was observed, as a result of microbial activity. It is always important to establish the best bioremediation method to be used in each particular site, since the success in a bioremediation process depends on environmental factors as well as on the composition of the microbiota present *in situ*.

Author details

Hilda Amelia Piñón-Castillo[1], Daniel Lardizabal Gutiérrez[2],
Francisco Javier Zavala-Díaz de la Serna[1], Daniel Hernández-Castillo[3],
Laila N. Muñoz-Castellanos[1], Blanca E. Rivera-Chavira[1] and
Guadalupe Virginia Nevárez-Moorillón[1]*

*Address all correspondence to: vnevare@uach.mx

1 Facultad de Ciencias Químicas, Universidad Autónoma de Chihuahua, Circuito Universitario s/n, Chihuahua, Chih., México

2 Centro de Investigación en Materiales Avanzados, Chihuahua, Chih., México

3 Universidad Anáhuac Norte, Naucalpan de Juárez, Méx., México

References

[1] FAO. Status of the World's Soil Resources (SWSR)—Main Report. Rome, Italy: Food and Agriculture Organization of the United Nations and Intergovernmental Technical Panel on Soils; 2015. 650 p

[2] Singh A, Parmar N, Kuhad RC, Ward OP. Bioaugmentation, biostimulation, and biocontrol in soil biology. In: Singh A, Parmar N, Kuhad RC, editors. Bioaugmentation, Biostimulation and Biocontrol. Berlin, Heidelberg: Springer; 2011. pp. 1-23. DOI: 10.1007/978-3-642-19769-7.ch1

[3] Cruz-Paredes C, Wallander H, Kjøller R, Rousk J. Using community trait-distributions to assign microbial responses to pH changes and Cd in forest soils treated with wood ash. Soil Biology and Biochemistry. 2017;**112**:153-164. DOI: 10.1016/j.soilbio.2017.05.004

[4] de Souza RB, Maziviero TG, Christofoletti CA, Pinheiro TG, Fontanetti CS. Soil contamination with heavy metals and petroleum derivates: Impact on edaphic fauna and remediation strategies. In: Hernandez Soriano MC, editor. Soil Processes and Current Trends in Quality Assessment. Rijeka, Croatia: InTech; 2013. pp. 175-203. DOI: 10.5772/52868

[5] Zawierucha I, Malina G. Bioremediation of contaminated soils: Effects of bioaugmentation and biostimulation on enhancing biodegradation of oil hydrocarbons. In: Singh A, Parmar N, Kuhad RC, editors. Bioaugmentation, Biostimulation and Biocontrol. Berlin, Heidelberg: Springer; 2011. pp. 187-202. DOI: 10.1007/978-3-642-19769-7 (ch 8)

[6] Vázquez-Luna D. Chronic toxicity of weathered oil-contaminated soil. In: Hernandez Soriano MC, editor. Environmental Risk Assessment of Soil Contamination. Rijeka, Croatia: InTech; 2014. pp. 87-103. DOI: 10.5772/57253

[7] Agnello AC, Bagard M, Van Hullebusch ED, Esposito G, Huguenot D. Comparative bioremediation of heavy metals and petroleum hydrocarbons co-contaminated soil by natural attenuation, phytoremediation, bioaugmentation and bioaugmentation-assisted

phytoremediation. Science of the Total Environment. 2016;**563**:693-703. DOI: 10.1016/j.scitotenv.2015.10.061

[8] Iturbe R, Flores-Serrano RM, Castro A, Flores C, Torres LG. Subsoil TPH contamination in two oil pipeline pumping stations and one pipeline right-of-way in north Mexico. Journal of Environmental Management. 2010;**91**(11):2396-2402. DOI: 10.1016/j.jenvman.2010.07.003

[9] Chiu HY, Hong A, Lin SL, Surampalli RY, Kao CM. Application of natural attenuation for the control of petroleum hydrocarbon plume: Mechanisms and effectiveness evaluation. Journal of Hydrology. 2013;**505**:126-137. DOI: 10.1016/j.jhydrol.2013.09.027

[10] Vásquez-Murrieta MS, Hernández-Hernández OJ, Cruz-Maya JA, Cancino-Díaz JC, Jan-Roblero J. Approaches for removal of PAHs in soils: Bioaugmentation, biostimulation and bioattenuation. In: Larramendy M, Solenski S, editors. Soil Contamination-Current Consequences and Further Solutions. Rijeka, Croatia: InTech; 2016. pp. 329-342. DOI: 10.5772/64682

[11] Abioye OP. Biological remediation of hydrocarbon and heavy metals contaminated soil. In: Pascucci S, editor. Soil Contamination. Rijeka, Croatia: InTech; 2011. pp. 127-142. DOI: 10.5772/24938

[12] Fontanetti CS, Nogarol LR, Souza RB, Perez DG, Maziviero GT. Bioindicators and biomarkers in the assessment of soil toxicity. In Pascucci S, editor. Soil contamination. Rijeka, Croatia: InTech; 2011. pp. 143-168. DOI: 10.5772/25042

[13] Das M, Adholeya A. Role of microorganisms in remediation of contaminated soil. In: Satyanarayana T, Johri BN, Prakash A, editors. Microorganisms in Environmental Management: Microbes and Environment. Dordrecht, Netherlands. Springer Science & Business Media; 2012. pp. 81-111. DOI: 10.1007/978-94-007-2229-3 (ch.4)

[14] Hazra C, Kundu D, Chaudhari A. Biosurfactant-assisted bioaugmentation in bioremediation. In: Satyanarayana T, Johri BN, Prakash A, editors. Microorganisms in Environmental Management: Microbes and Environment. Dordrecht, Netherlands. Springer Science & Business Media; 2012. pp. 631-664. DOI 10.1007/978-94-007-2229-3 (ch. 28)

[15] Kuppusamy S, Thavamani P, Venkateswarlu K, Lee YB, Naidu R, Megharaj M. Remediation approaches for polycyclic aromatic hydrocarbons (PAHs) contaminated soils: Technological constraints, emerging trends and future directions. Chemosphere. 2017;**168**:944-968. DOI: 10.1016/j.chemosphere.2016.10.115

[16] Martinkosky L, Barkley J, Sabadell G, Gough H, Davidson S. Earthworms (*Eisenia fetida*) demonstrate potential for use in soil bioremediation by increasing the degradation rates of heavy crude oil hydrocarbons. Science of the Total Environment. 2017;**580**:734-743. DOI: 10.1016/j.scitotenv.2016.12.020

[17] Couto MN, Monteiro E, Vasconcelos MT. Mesocosm trials of bioremediation of contaminated soil of a petroleum refinery: Comparison of natural attenuation, biostimulation and bioaugmentation. Environmental Science and Pollution Research. 2010;**17**(7):1339-1346. DOI: 10.1007/s11356-010-0318-y

[18] Montenegro IP, Mucha AP, Reis I, Rodrigues P, Almeida CM. Copper effect in petroleum hydrocarbons biodegradation by microorganisms associated to *Juncus maritimus*: Role of autochthonous bioaugmentation. International Journal of Environmental Science and Technology. 2017;**14**(5):943-955. DOI: 10.1007/s13762-016-1215-9

[19] Silva-Castro GA, Rodriguez-Calvo A, Laguna J, González-López J, Calvo C. Autochthonous microbial responses and hydrocarbons degradation in polluted soil during biostimulating treatments under different soil moisture. Assay in pilot plant. International Biodeterioration & Biodegradation. 2016;**108**:91-98. DOI: 10.1016/j.ibiod.2015.12.009

[20] Gomez F, Sartaj M. Optimization of field scale biopiles for bioremediation of petroleum hydrocarbon contaminated soil at low temperature conditions by response surface methodology (RSM). International Biodeterioration & Biodegradation. 2014;**89**:103-109. DOI: 10.1016/j.ibiod.2014.01.010

[21] Chen M, Xu P, Zeng G, Yang C, Huang D, Zhang J. Bioremediation of soils contaminated with polycyclic aromatic hydrocarbons, petroleum, pesticides, chlorophenols and heavy metals by composting: Applications, microbes and future research needs. Biotechnology Advances. 2015;**33**(6):745-755. DOI: 10.1016/j.biotechadv.2015.05.003

[22] Chagas-Spinelli AC, Kato MT, de Lima ES, Gavazza S. Bioremediation of a tropical clay soil contaminated with diesel oil. Journal of Environmental Management. 2012;**113**:510-516. DOI: 10.1016/j.jenvman.2012.05.027

[23] Nevárez-Moorillón GV, Piñón-Castillo H, Torres-Muñoz V, Muñoz-Castellanos LN, Vélez-Sánchez Verín C, Espinoza JV, Riojas-González HH, Hernández-Castillo D. Soil contaminated with combustoleo (residual fuel no. 6). Remediation strategies. In: Magar VS, Kelley ME, editors. In Situ and On-site Bioremediation-2003. Proceedings of the Seventh International In Situ and On Site Bioremediation Symposium (Orlando, Fl.; June 2003). Columbia, OH, USA; Battelle Press: 2003. pp. E-22

[24] Alzaga R, Montuori P, Ortiz L, Bayona JM, Albaigés J. Fast solid-phase extraction–gas chromatography–mass spectrometry procedure for oil fingerprinting: Application to the Prestige oil spill. Journal of Chromatography A. 2004;**1025**(1):133-138. DOI: 10.1016/j.chroma.2003.10.080

[25] Gallego JR, González-Rojas E, Peláez AI, Sánchez J, García-Martínez MJ, Ortiz JE, Torres T, Llamas JF. Natural attenuation and bioremediation of Prestige fuel oil along the Atlantic coast of Galicia (Spain). Organic Geochemistry. 2006;**37**(12):1869-1884. DOI: 10.1016/j.orggeochem.2006.07.022

[26] Kiyohara H, Nagao K, Yano K. Rapid screen for bacteria degrading water-insoluble, solid hydrocarbons on agar plates. Applied and Environmental Microbiology. 1982;**43**(2):454-457

[27] Langworthy DE, Stapleton RD, Sayler GS, Findlay RH. Genotypic and phenotypic responses of a riverine microbial community to polycyclic aromatic hydrocarbon contamination. Applied and Environmental Microbiology. 1998;**64**(9):3422-3428

[28] Environmental Protection Agency (EPA). Methods for chemical analysis of water and wastes. EPA 8440. Cincinnati, OH: U.S. Environmental Protection Agency. Environmental Monitoring and Support Laboratory; 1996

[29] Environmental Protection Agency (EPA). Methods for chemical analysis of water and wastes. EPA 3570. Cincinnati, OH: U.S. Environmental Protection Agency. Environmental Monitoring and Support Laboratory; 2002

[30] Gubry-Rangin C, Novotnik B, Mandič-Mulec I, Nicol GW, Prosser JI. Temperature responses of soil ammonia-oxidising archaea depend on pH. Soil Biology and Biochemistry. 2017;**106**: 61-68. DOI: 10.1016/j.soilbio.2016.12.007

[31] Fernandez-Luquentilde F, Vazquez-Nuntilde E, Martinez-Suarez C, Salomon-Hernandez G, Valenzuela-Encinas C, Hernandez OF, Ceballos-Ramirez JM, Dendooven L. Bacterial community composition of a saline-alkaline soil amended or not with wastewater sludge and contaminated with polycyclic aromatic hydrocarbons (PAHs). African Journal of Microbiology Research. 2013;**7**(28):3605-3614. DOI: 10.5897/AJMR2012.2327

[32] Martínez-Prado A, Pérez-López M, Pinto-Espinoza J, Gurrola-Nevárez BA, Osorio-Rodríguez AL. Biorremediación de suelo contaminado con hidrocarburos empleando lodos residuales como fuente alterna de nutrientes. Revista internacional de contaminación ambiental. 2011;**27**(3):241-252

[33] Taccari M, Milanovic V, Comitini F, Casucci C, Ciani M. Effects of biostimulation and bioaugmentation on diesel removal and bacterial community. International Biodeterioration & Biodegradation. 2012;**66**(1):39-46. DOI: 10.1016/j.ibiod.2011.09.012

[34] Varjani SJ. Microbial degradation of petroleum hydrocarbons. Bioresource Technology. 2017;**223**:277-286. DOI: 10.1016/j.biortech.2016.10.037

[35] Chebbi A, Hentati D, Zaghden H, Baccar N, Rezgui F, Chalbi M, Sayadi S, Chamkha M. Polycyclic aromatic hydrocarbon degradation and biosurfactant production by a newly isolated *Pseudomonas* sp. strain from used motor oil-contaminated soil. International Biodeterioration & Biodegradation. 2017;**122**:128-140. DOI: 10.1016/j.ibiod.2017.05.006

[36] Di Gregorio S, Siracusa G, Becarelli S, Mariotti L, Gentini A, Lorenzi R. Isolation and characterization of a hydrocarbonoclastic bacterial enrichment from total petroleum hydrocarbon contaminated sediments: Potential candidates for bioaugmentation in bio-based processes. Environmental Science and Pollution Research. 2016;**23**(11):10587-10594. DOI: 10.1007/s11356-015-5944-y

[37] Sousa ST, Cabral L, Júnior L, Vieira G, Oliveira VM. Diversity of aromatic hydroxylating dioxygenase genes in mangrove microbiome and their biogeographic patterns across global sites. MicrobiologyOpen. 2017;e490. DOI: 10.1002/mbo3.490

[38] Sarkar P, Roy A, Pal S, Mohapatra B, Kazy SK, Maiti MK, Sar P. Enrichment and characterization of hydrocarbon-degrading bacteria from petroleum refinery waste as

potent bioaugmentation agent for in situ bioremediation. Bioresource Technology. DOI: 10.1016/j.biortech.2017.05.010

[39] Vaidya S, Jain K, Madamwar D. Metabolism of pyrene through phthalic acid pathway by enriched bacterial consortium composed of *Pseudomonas*, *Burkholderia*, and *Rhodococcus* (PBR). 3 Biotech. 2017;**7**(1):29. DOI: 10.1007/s13205-017-0598-8

[40] Roy AS, Baruah R, Borah M, Singh AK, Boruah HP, Saikia N, Deka M, Dutta N, Bora TC. Bioremediation potential of native hydrocarbon degrading bacterial strains in crude oil contaminated soil under microcosm study. International Biodeterioration & Biodegradation. 2014;**94**:79-89. DOI: 10.1016/j.ibiod.2014.03.024

[41] Viramontes-Ramos S, Portillo-Ruiz MC, Ballinas-Casarrubias MD, Torres-Muñoz JV, Rivera-Chavira BE, Nevárez-Moorillón GV. Selection of biosurfactan/bioemulsifier-producing bacteria from hydrocarbon-contaminated soil. Brazilian Journal of Microbiology. 2010;**41**(3):668-675. DOI: 10.1590/S1517-83822010000300017

[42] Mudgil P. Biosurfactants for soil biology. In: Singh A, Parmar N, Kuhad RC, editors. Bioaugmentation, Biostimulation and Biocontrol. Berlin, Heidelberg; Springer; 2011. pp. 203-220. DOI: 10.1007/978-3-642-19769-7 (ch. 9)

[43] Diario Oficial de la Federacion. Norma oficial mexicana de emergencia NOM-EM-138-ECOL-2002. Que establece los límites máximos permisibles de contaminación en suelos afectados por hidrocarburos, la caracterización del sitio y procedimientos para la restauración. 2002

[44] Guarino C, Spada V, Sciarrillo R. Assessment of three approaches of bioremediation (Natural Attenuation, Landfarming and Bioagumentation–Assistited Landfarming) for a petroleum hydrocarbons contaminated soil. Chemosphere. 2017;**170**:10-16. DOI: 10.1016/j.chemosphere.2016.11.165

[45] Liu PW, Chang TC, Chen CH, Wang MZ, Hsu HW. Bioaugmentation efficiency investigation on soil organic matters and microbial community shift of diesel-contaminated soils. International Biodeterioration & Biodegradation. 2014;**95**:276-284. DOI: 10.1016/j.ibiod.2014.05.004

[46] White JC, Kelsey JW, Hatzinger PB, Alexander M. Factors affecting sequestration and bioavailability of phenanthrene in soils. Environmental Toxicology and Chemistry. 1997;**16**(10):2040-2045. DOI: 10.1002/etc.5620161008

[47] Yang Y, Shu L, Wang X, Xing B, Tao S. Mechanisms regulating bioavailability of phenanthrene sorbed on a peat soil-origin humic substance. Environmental Toxicology and Chemistry. 2012;**31**(7):1431-1437. DOI: 10.1002/etc.1844

[48] Wu M, Li W, Dick WA, Ye X, Chen K, Kost D, Chen L. Bioremediation of hydrocarbon degradation in a petroleum-contaminated soil and microbial population and activity determination. Chemosphere. 2017;**169**:124-130. DOI: 10.1016/j.chemosphere.2016.11.059

[49] Lladó S, Solanas AM, De Lapuente J, Borras M, Viñas M. A diversified approach to evaluate biostimulation and bioaugmentation strategies for heavy-oil-contaminated soil. Science of the Total Environment. 2012;**435**:262-269. DOI: 10.1016/j.scitotenv.2012.07.032

[50] Koolivand A, Rajaei MS, Ghanadzadeh MJ, Saeedi R, Abtahi H, Godini K. Bioremediation of storage tank bottom sludge by using a two-stage composting system: Effect of mixing ratio and nutrients addition. Bioresource Technology. 2017;**235**:240-249. DOI: 10.1016/j.biortech.2017.03.100

Effectiveness of Sorghum Husk and Chicken Manure in Bioremediation of Crude Oil Contaminated Soil

Feyisayo V. Adams, Maryam F. Awode and
Bolade O. Agboola

Additional information is available at the end of the chapter

http://dx.doi.org/10.5772/intechopen.71832

Abstract

This study identifies total petroleum hydrocarbons (TPH) and traces of heavy metals such as zinc, lead, cadmium, nickel, and copper in crude oil-contaminated soil. It also focuses on the use of poultry manure and sorghum husk in the bioremediation of the contaminated soil. Crude oil-contaminated soil sample was divided into five parts (A: untreated, B: poultry manure, C and D: poultry manure and sorghum husk in ratios 1:1 and 3:1, respectively, and E: sorghum husk). The heavy metals concentrations and TPH content were assessed initially in the untreated soil sample and later on the 5th, 10th, 15th, 20th, 25th, and 30th days after adding the stimulants. Gas chromatography-mass spectroscopy (GC-MS), atomic absorption spectrometer (AAS), pH and conductivity meters were used for TPH, heavy metals, pH and electrical conductivity analyses, respectively. The results showed soil sample C to have highest TPH reduction, while the soil sample E exhibited 96.1% reduction in nickel, 97.5% reduction in zinc, 100% reduction in lead, and 99.3% reduction in copper. The pH of the soil ranged from 7.13 to 7.92 (within the range 6.5–8 suitable for microbial growth). The electrical conductivity for soil samples B–E increased and also in the acceptable range of 130–2320 μS/cm.

Keywords: bioremediation, crude oil, contaminated soil, sorghum husk, poultry manure

1. Introduction

Bioremediation is a cost effective and environment friendly method, which involves the use of microorganisms to degrade contaminants from soil and water. It is the use of naturally cropped up microorganisms by humans in an attempt to decontaminate man-made contaminants. The advantage of this technique is that it is conservative of soil characteristic and texture, but the success of its application is dependent on the nature of the contaminant in the soil and environmental conditions under which the test will be carried out. Because bioremediation utilizes microbial activity, it is therefore necessary to provide nutrients, oxygen, and suitable temperature to maximize growth. It has been reported that the use of organic waste enhances bioremediation of crude oil-polluted soil by facilitating soil aeration and raising the water holding ability of the soil [1].

Bioremediation promotes natural clean-up of dangerous toxins by the environment and can either be done in-situ or ex-situ. In-situ bioremediation is carried out at the site of interest whereas in ex-situ bioremediation, the contaminated soil is collected and processed at an offsite area such as a laboratory for cleansing [2]. The main procedure frequently used in oil-spill bioremediation technique is biostimulation, that is, through the addition of nutrients (nitrogen, phosphorus, and potassium) and regulation of environmental conditions to maximize growth of native bacteria and biological degradation. Thus, bioremediation can be said to be a method, which utilizes microbial diversity and metabolic flexibility to convert chemical contaminants into less toxic products [3]. Various studies have investigated the part of microorganisms and bulking agents such as rice husk in the remediation of oil-contaminated soils [4, 5].

A research by Hidayah and Mangkoedihardjo [5] on remediation of chromium-contaminated soil using rice husk showed that microbial growth of indigenous bacteria increased by three folds with a corresponding fall in chromium levels in the first 2 weeks. The treatments adopted comprised of soil and rice husk in varied ratios. The results proved that rice husk helps in biostimulation and has a significant role in improving bioremediation of chromium-polluted soil [5]. Additionally, the study by Adams et al. [4] on the use of rice husk, chicken manure, and their combination for bioremediation of crude oil-contaminated soil revealed that rice husk removed more total petroleum hydrocarbons (TPH) compared to the chicken manure and their combination.

Other agricultural waste that has been used for crude oil-contaminated soil bioremediation includes melon husk [6] and peanut hulls [7]. Sorghum husk moreover is also known to contain some nutrients that can enhance microbial growth and also adsorb heavy metals. Red sorghum husk is a rich source of anthocyanins, which are plant pigments known to increase microbial activity. It is low in protein and ash content but rich in fibrous materials.

Ogboghodo et al. [8] stated that the addition of poultry manure to crude oil-polluted soil caused an increase in plant height from 20 to 149 cm as level of manure applied increased from 0 to 150 kg/ha and dry matter yield also increased from 27 to 58 g within 2 weeks of study, and proposed the adoption of poultry manure for the stimulation of hydrocarbon in the soil as a good technique of battling petroleum contamination in the natural environment. Another study by Okafor et al. [9] has proven that poultry manure is rich in organic matter thus allows for the growth of a substantial amount of microorganisms. Varied concentrations of poultry manure were studied and a positive correlation between growth of microorganism and increase in poultry droppings was observed. The sample amended with the highest amount of poultry droppings exhibited greatest reduction in TPH. Isolates of the hydrocarbon degrading microorganisms in this study include Bacillus spp., Pseudomonas

http://dx.doi.org/10.5772/intechopen.71832

spp., Flavobacterium spp., Fusarium spp., and Aspergillus spp., while the total heterotrophic bacterial and fungal counts of the poultry droppings were 4.2×10^4 and 1.8×10^4 cfu/g [9]. No work has been done on the study of the effectiveness of sorghum husk and its combination with poultry manure in bioremediation of contaminated soil. Thus, this present work intends to study the effectiveness of the sorghum husk, poultry manure, and their combination in bioremediation of crude oil-contaminated soil from the Niger-Delta Nigeria.

2. Materials and methods

2.1. Materials

The crude oil-contaminated soil used in this study was obtained from the Niger-Delta region of Nigeria. A total 15 plates (5 samples in triplicates; see **Table 1**) were used as simulated environment, while, pH meter, conductivity meter, GC-MS, and atomic absorption spectrometry (AAS) were used to determine the soil pH, electrical conductivity, TPH, and heavy metals, respectively. Other devices used were electronic weighing balance conical flasks, beakers, spatula, and stirrers. Hexane was used as solvent for the extraction purpose for all the analyses that were carried out.

2.2. Soil samples preparation

The soil samples were prepared for bioremediation by removing sticks and stones after which the different stimulants (sorghum husk, chicken manure, and their mixtures) were then added in varied amounts to enhance hydrocarbon degradation and successful microbial growth. About 200 g of the crude oil-contaminated soil was weighed into each of the plastic containers and labeled accordingly (see **Table 1**) in triplicates. The soil sample used was divided into five parts (A: untreated, B: poultry manure only, C and D: poultry manure and sorghum husk in ratios 1:1 and 3:1, respectively, and E: sorghum husk only). In order to avoid low oxygen and preparing aerobic soil conditions, the soil samples were mixed twice per week for a period of 30 days. The soil samples were kept under a controlled temperature of 29°C and their pH were kept in a range of 6.5–8.5, which is the optimal range for microbial growth and reproducibility [4].

Soil samples	Description
A	Contaminated soil only
B	Contaminated soil mixed with poultry manure only (8/200 g-soil)
C	Contaminated soil mixed with poultry manure and sorghum husk in the ratio 1:1 (16/200 g-soil)
D	Contaminated soil mixed with poultry manure and sorghum husk in the ratio 3:1 (32/200 g-soil)
E	Contaminated soil mixed with sorghum husk only (8/200 g-soil)

Table 1. Soil sample preparations.

2.3. pH and conductivity

The samples for pH and conductivity determination were prepared by adding 5 g of the soil sample in 10 ml deionized water. The mixture was stirred and kept for 30 minutes before the conductivity and pH soil were determined. The first values were obtained on the 5th day of the study and thus served as the initial pH of the soil samples.

2.4. Total petroleum hydrocarbon and heavy metals determination

An extraction process was carried out by putting 20 g of soil sample in 20 ml of hexane and shaking it mechanically for 30 minutes. The mixture was then placed in a centrifuge to separate the liquid component from the solid component. The liquid component was obtained from the mixture by filtration and the filtrates were analyzed using GC-MS (this process was repeated every 5 days for 15 days) and AAS for TPH and heavy metal reduction, respectively. The heavy metals in the soil samples were determined after the 30 days of study.

2.5. Total organic carbon (TOC)

There are broad ranges of organic carbon in soil from the breakdown of either plants or animals [10]. TOC is useful in assessing quantity of organic matter in a soil sample. The Walkley-Black chromic acid wet oxidation method as described by Resources [11] was used in determining the TOC in the soil samples.

$$\mathrm{TOC}(\%) = \frac{0.003\ \mathrm{g} \times \mathrm{N} \times 10\ \mathrm{ml} \times (1 - T/S) \times 100}{\mathrm{ODW}} \tag{1}$$

where N = concentration of $K_2Cr_2O_7$ solution, T = Volume of $FeSO_4$ used in sample titration (mL), S = Volume of $FeSO_4$ used in blank titration (mL), ODW = Oven-dry sample weight (g).

3. Results and discussion

3.1. pH and conductivity

The pH of the soil is a significant soil parameter. **Figure 1** shows the changes in the pH values for all the soil samples with time. The pH values were increasing and decreasing in an irregular manner. Soil pH can be highly variable depending on the type of soil and location of the soil. The soil pH ranges from 2.5 in mine spoils to 11 in alkaline deserts [12]. Extreme pH of soils would have a negative influence on the ability of microbial populations to degrade hydrocarbons. Soil sample A showed the highest pH values from the 5 to 20th day (7.55, 7.56, 7.70, and 7.61, respectively), while the soil sample B exhibited the lowest pH value on the 5th day, and E on the 10th, 15th, and 20th day, respectively. The pH values of the samples containing stimulants (samples B–E) had lower pH values compared to the control sample (sample A) in the first 20 days of the study. However, on the 25th and 30th day, sample B exhibited the highest pH value contrary to its pH (lowest) in the first 5 days of the study.

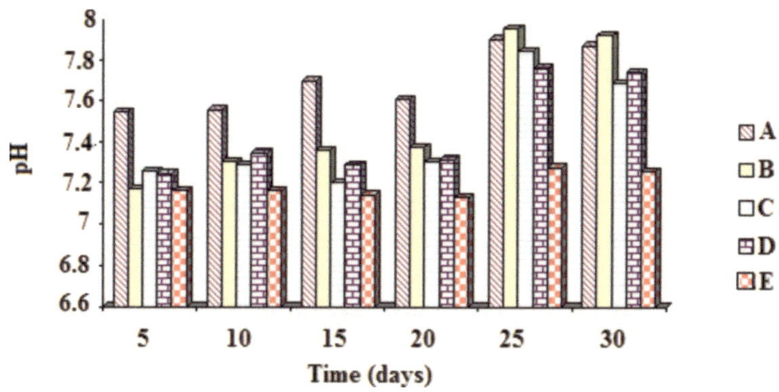

Figure 1. Change in pH with time.

Addition of water after day 10 caused a rapid increase in the pH values for samples A and B, while others decreased. However, pH value continued to increase till the end of the study for all the soil samples. The change in pH could be attributed to increase in metabolic activities of microbes and the degradation of hydrocarbons in the soil that may have led to the synthesis of intermediary metabolites [13]. It is important to note that the pH was within proper range (6–8) for microbial growth throughout the treatment process [14].

Figure 2 shows the electrical conductivity (EC) results obtained for the soil samples from day 5 to 30 of the bioremediation study. The electric conductivity ranges between 159 and 1533 μS/cm for all the soil samples. It was observed that the conductivity initially increased in the first 15 days before decreasing in the last 20–30 days. The soil sample B showed highest electrical conductivity throughout the study period. The soil samples C and D also exhibited higher EC compared to the soil sample E. The control soil sample A had the lowest EC throughout the study period. The increase can be due to the presence of nutrients at the beginning of the treatment, which was gradually decreased with time by microbes leading to the observed decrease.

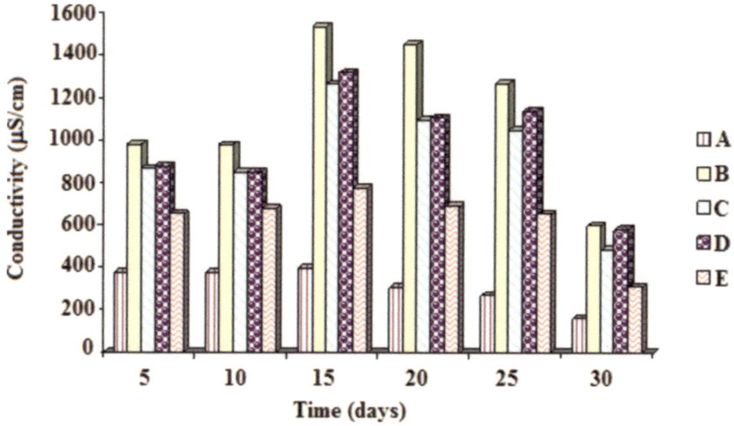

Figure 2. Change in electric conductivity with time.

Correlating the pH with the EC results (**Figures 1** and **2**) showed that at the first 15 days of the study, the pH values were generally lower with higher EC values. Also, at the last 15 days, higher values of pH were observed with lower EC values. These results indicate higher microbial activities in the last 15 days and possible higher TPH degradation, which lowered the EC of the soil samples thereby increasing the pH values. It is known that EC depends on the concentration of all the ions present in the soil samples. The higher the concentration of the ions in the soil, the stronger or higher the EC values for the soil. Soil that exhibits high alkalinity will have less soluble salt in it [15], which indicates that at low soil pH value, high soluble salt content is present in the soil. Thus, the soil will have high EC.

3.2. Heavy metals

The concentration of heavy metals in the soil samples was determined using AAS. **Table 2** shows the percentage removal of heavy metals from the crude oil-contaminated soil. The initial contaminated soil had traces of copper, nickel, zinc, and lead (0.1840, 0.0820, 0.4120, and 0.0198 mg/g, respectively). A significant reduction in the amount of heavy metals in all the soil samples was observed after a 30-day bioremediation study. The decrease in the heavy metal in soil sample A could be due to moisture, aeration, and natural bioremediation due to the natural microbes in the soil sample. Sorghum husk (soil sample E) showed more heavy metal removal efficiency compared to poultry manure (soil sample B) and their mixtures (soil samples C and D). The highest removal of >99% was observed in samples A, C, D, and E for Cu, while the lowest metal removal was for Pb in soil samples A, B, and D. However, Pb was totally removed from soil sample E, which contained sorghum husk only and about 94.9% of Pb was removed from sample C, which contained a mixture of poultry manure and sorghum husk in ratio 1:1.

Metal removal in soil sample A indicates that microorganisms indigenous to that soil sample have greater metal reduction ability than those present in soil samples C, B, and D to which stimulants was added. Nonliving biomass such as rice husks and corn cobs have been proven to be effective in adsorption [4, 5, 16]. The results obtained in this study show that sorghum husk also has high capacity for adsorption of heavy metals due to its high fiber [17, 18].

Heavy metals	A (%)	B (%)	C (%)	D (%)	E (%)
Ni	95.5	94.6	95.1	93.7	96.1
Zn	97.3	96.0	96.8	92.5	97.5
Pb	69.7	82.8	94.9	76.8	100.0
Cu	99.2	95.0	99.3	99.1	99.3
TOC	0.08	0.78	0.70	0.48	0.74

Table 2. Removal ratios of heavy metals from the soil samples and TOC in the soil.

3.3. Total organic carbon

The results obtained from the total organic carbon analysis are presented in **Table 2**. The TOC of the untreated soil sample was 0.15%. A reduction in TOC is observed only in the control soil sample A. The reduction of TOC in A indicates natural microbial activity in the soil since microorganisms tend to use soil organic carbon as a source of energy. Additionally, the increase in TOC in the other groups can be attributed to the decomposition of the plant and animal wastes that were added.

3.4. Total petroleum hydrocarbon

The results from the GC-MS analysis showing the spectra for all the untreated and the treated soil samples on day 5 and 15 are shown in **Figures 3** and **4** and the major TPH present in the soil samples at those days are presented in **Tables 3** and **4**.

The results showed about 23 TPH present in the crude oil-contaminated soil. However, the control soil sample showed a reduction to about six major TPH on the 5th day and only four major ones on the 15th day (**Table 4**). This could be due to the indigenous microbial in the soil, which have the ability to generally breakdown heavy metals and petroleum hydrocarbons as long as favorable environmental condition is provided for them to carry out their metabolic actions.

The benefits of relying on these microorganisms for biodegradation instead of adding new ones are that they are attuned for survival and growth in that condition. Furthermore, their capacity to use hydrocarbons is dispersed among an assorted microbial populace. These populaces take place in common biological systems and either autonomously or jointly process different hydrocarbons [19, 20].

The results of the control sample show biodegradation of the TPH by natural microorganisms present in the soil. Constant watering and aeration of samples after the 10th day enhanced metabolic microbial activity on the 15th days, but the absence of stimulants in the treatment groups may have contributed to slow degradation of hydrocarbons. Biodegradability of hydrocarbons in soil has shown correlation to their water solubility [17]. This is because bacteria in the unsaturated soil occur mainly in the interstitial water of soil. Therefore, solubility of the chemical will determine its concentration in soil. The control group showed high concentrations of 9,12-octadecadienoic acid, ethyl oleate and tetracosane, which were absent in the original sample. A reduction in the amount of 1-nonadecene is noticed from 12.20 s in the untreated sample to 11.23 s in control group A. As generally observed in the degradation of TPH results, it can be seen that the degradation was faster (15 days). This was due to the high temperature (>30°C) at the time of the study in Yola (North Eastern, Nigeria). Temperature affects the solubility of TPH [21]. Hydrocarbon biodegradation can however take place over a wide ranges of temperature (highest degradation rates usually take place between 30 and 40°C in soil environments, 15–20°C in marine environments, and 20–30°C in some fresh water environments [22, 23].

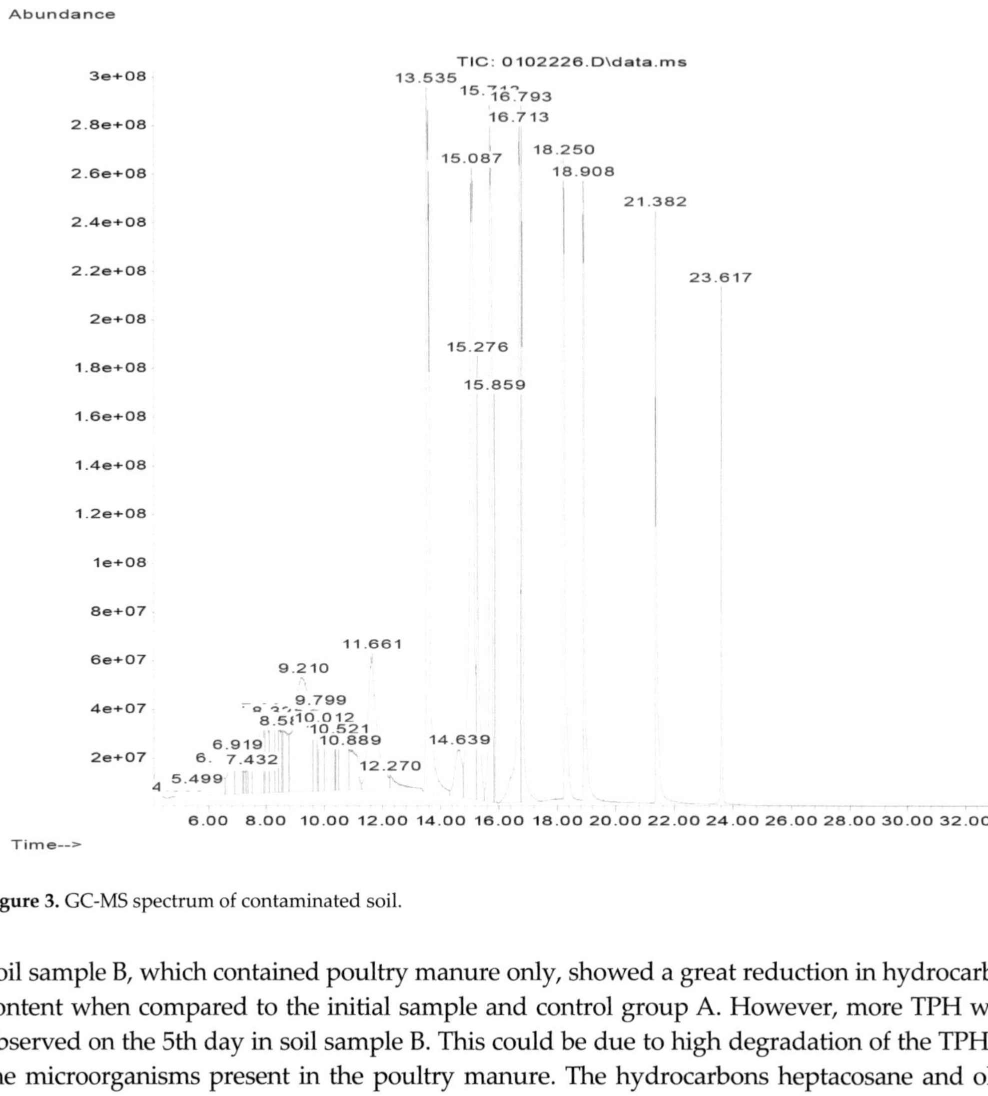

Figure 3. GC-MS spectrum of contaminated soil.

Soil sample B, which contained poultry manure only, showed a great reduction in hydrocarbon content when compared to the initial sample and control group A. However, more TPH were observed on the 5th day in soil sample B. This could be due to high degradation of the TPH by the microorganisms present in the poultry manure. The hydrocarbons heptacosane and oleic acid were present in high concentrations of 58.44 and 40.71, respectively (**Table 4**). Formation of oleic acid serves as the evidence of oxidation. The biodegradation of oil pollutants in soil needs necessary nutrients. Choi et al. [24], Kim et al. [25], and Pelletier et al. [26] studied the effect of fertilizers on crude oil-contaminated soil bioremediation in sub-Antarctic intertidal sediments over a 1 year. The results of the study showed that microbial, chemical, and toxicological parameters demonstrated the use of various fertilizers in a pristine environment. In study using poultry manure as organic fertilizer in contaminated soil increased biodegradation was reported but the extent of biodegradation was influenced by the incorporation of alternate carbon substrates or surfactants [27].

It can also be seen in **Table 4** and **Figure 4**, there is a reduction in the hydrocarbon composition of the soil sample C in comparison to the initial untreated soil sample. Concentrations of 4-heptenal, 1-heptene, octadecane, 1-chloro-acetamide, and methyl 7-oxopentadecanoate were quite low relative to that of octadecanal, heptacosane, hexadecanoic acid, and methyl ester in the

http://dx.doi.org/10.5772/intechopen.71832

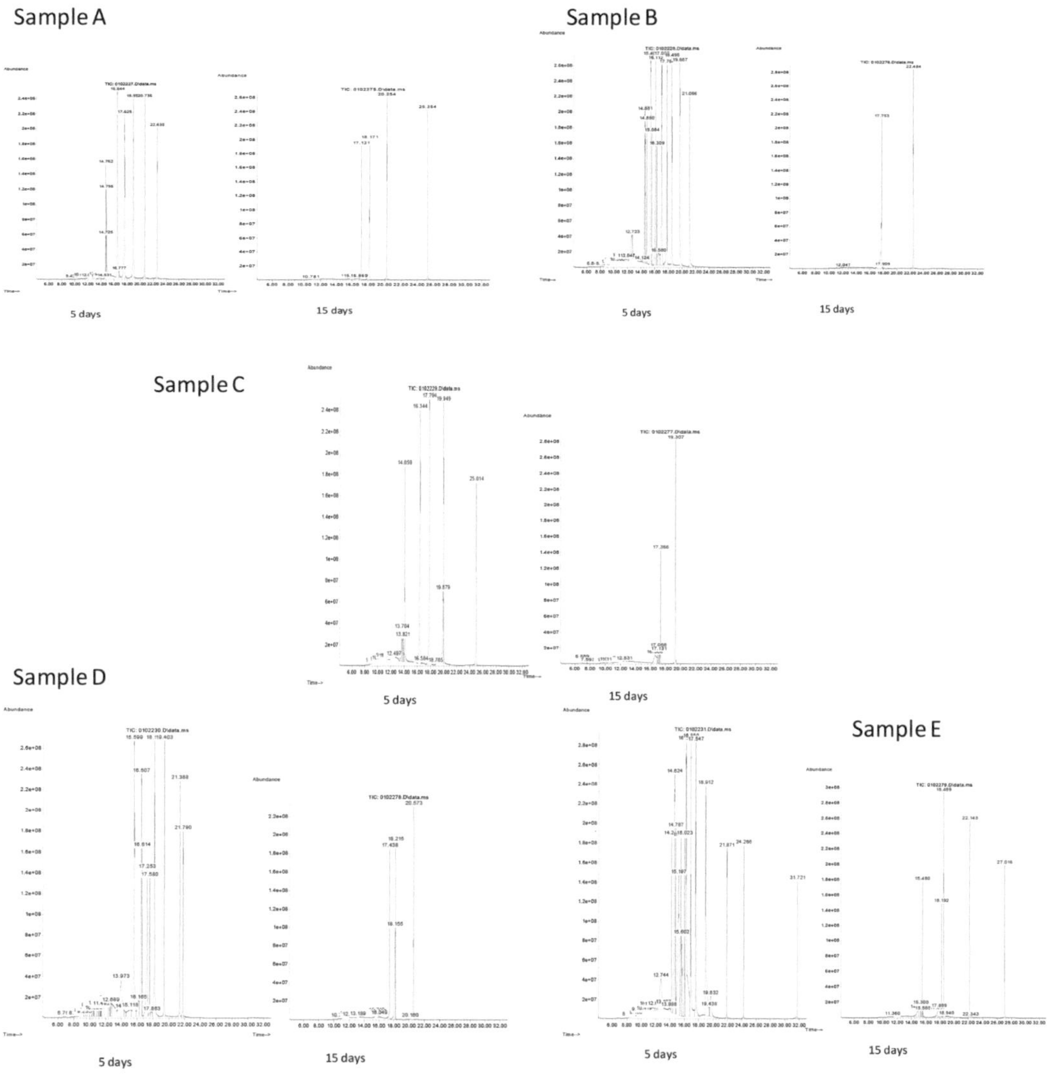

Figure 4. GC-MS spectra of soil samples A–E.

contaminated soil sample. The presence of esters and other oxidized compounds is an indication that the presence of poultry droppings and sorghum husk induced oxidation of the hydrocarbon compounds. The presence of hexadecanoic acid in soil samples C and D is as a result of degradation by sorghum husk (**Table 4**) and oleic acid by the poultry manure. The carbon chains are also shorter ranging from C_7 to C_{20}. The microbial attack on hydrocarbons depends on the type of hydrocarbon. The hydrocarbons susceptibility to microbial attack is ranked in the following order: n-alkanes > branched alkanes > low molecular weight aromatics > cyclic alkanes [28].

Table 4 shows that treatment of contaminated soil with poultry manure and sorghum husk in a 3:1 ratio (sample D) resulted in a decrease in hydrocarbon components. There was oxidation of some of the compounds from the initial crude oil-contaminated soil analysis to 1-hexanoic acid and 1,2,15-pentadecanetriol.

No	Retention time (min)	Area (%)	Major molecule
1	6.356	1.13	Heptane
2	6.839	1.08	Nonadecanol
3	7.896	2.16	Dodecane
4	7.967	1.16	Oxirane
5	8.266	1.49	Octadecane, 1-Chloro-acetamide
6	8.325	1.03	Cyclohexane
7	8.582	1.45	17-Pentatriacontene
8	9.209	8.41	Hexadecane
9	9.674	1.44	Silane, trichloroeicosylsilane
10	9.800	1.87	Dodecane, 1-Choro-acetamide
11	10.522	2.01	1-Eicosanol
12	10.886	1.44	Cyclohexane
13	11.662	5.46	Pentalene
14	13.536	10.56	Cyclotetradecane
15	14.640	1.56	Cyclohexanol
16	15.088	12.20	1-Nonadecene
17	15.273	3.61	1-Docosanethiol
18	15.715	7.95	Cyclopentadecane
19	16.712	8.60	5-Eicosene
20	16.795	4.26	1-Acetyl-3-(2-pyridyl)-4-phenylpyrazoline
21	18.251	4.19	9-Hexacosene
22	18.908	4.26	Tricosane
23	21.385	3.90	4-Hexenoic acid, 3-Methyl-2,6-dioxo-4-hexenoic acid

Table 3. Major composition of untreated soil sample.

Table 4 and **Figure 4** show a reduction in the number of TPH in the soil sample E compared to that in the initial untreated soil sample. There was an addition of chloride to most of the hydrocarbon compounds in this soil sample; this could be due to the presence of the sorghum husk. The soil sample E showed the highest amount of hydrocarbon compounds in comparison to other treatment groups on the 5th and 15th days. This could be due to the pH (lowest) of the soil sample on these days as compared to the other soil samples. Reports have shown that higher biodegradation is achieved at increased pH values. According to Verstraete et al. [29], adjusting the pH value of gasoline from more acidic (4.5) to neutral (7.4) doubled its rate of biodegradation. Meanwhile, the rates decreased significantly when the pH was raised again to 8.5. Also, the optimum pH recorded by Dibble and Bartha [30] for the mineralization of oily sludge in soil ranged from 5.0 to 7.8.

Samples	Days	Retention time (min)	Area (%)	Main probable molecules
A	5	14.762	3.63	11-Hexacosyne
		14.795	7.56	Ergosta-8,14-dien-3-ol
		16.544	25.43	Oxirane
		17.625	25.27	1-Octadecene
		18.991	16.12	Oxirane
		20.736	9.48	Octadecane, 1-(ethenyloxy)-octadecane
		22.638	6.96	Ethyl Oleate
	15	15.661	1.32	Hexane
		15.894	1.45	Undecane
		15.983	1.83	3,3′-Oxybis-azocine
		17.129	11.23	1-Nonadecene
		18.174	37.38	9,12-Octadecadienoic acid
		20.257	22.32	Ethyl Oleate
		25.253	23.69	Tetracosane
B	5	14.533	5.65	Cyclohexadecane
		15.404	13.95	1-Nonadecene
		16.115	13.32	Oxirane
		17.763	11.02	Ethyl Oleate
		18.496	5.35	2-Bromo-6-methyl heptane
		19.654	6.64	Tetracosane
		21.069	6.64	15-Tetracosenoic acid, Methyl ester
	15	17.756	58.44	Heptacosane
		22.483	40.71	Oleic acid
C	5	13.821	2.64	Tridecane
		14.062	9.25	9-Undecen-2-one
		16.344	27.95	Oxirane
		17.794	23.48	1-Octadecene
		19.879	6.73	9-Octadecenoic acid, Methyl ester
		19.949	14.29	9-Hexacosene
		25.014	6.25	Phemetrazine
	15	6.654	1.58	4-Heptenal
		6.887	1.78	1-Heptene
		16.586	14.13	Octadecanal
		16.986	7.83	Oleic acid
		17.070	5.85	Octadecane, 1-Chloro-acetamide
		17.129	5.66	Methyl 7-oxopentadecanoate
		17.368	12.92	Heptacosane
		19.308	43.48	Hexadecanoic acid, Methyl ester

Samples	Days	Retention time (min)	Area (%)	Main probable molecules
D	5	13.972	7.69	10-Octadecenal
		15.601	17.87	3-Eicosene
		16.507	14.51	1-Octadecene
		17.254	10.28	1-Docosene
		18.165	9.98	9-Octadecene
		19.403	5.51	1-Nonadecene
		21.790	3.90	Heneicosane
	15	15.714	4.69	1-hexanoic acid
		16.049	1.00	1,2,15-Pentadecanetriol
		17.439	16.63	Tetracosane
		18.156	31.11	Eicosane
		18.215	12.73	Tricosane
		20.573	30.98	9-Hexacosene
E	5	12.744	4.75	10-Octadecenal
		14.624	11.31	1-Nonadecene
		15.197	12.71	Hexadecanoic acid, Methyl ester
		15.603	4.87	Pentadecanoic acid, Methyl ester
		16.198	21.27	Heneicosane
		16.850	9.27	12-Octadecenoic acid, Methyl ester
		18.912	9.70	9-Hexacosene
		21.871	2.08	Hexadecanol
		24.266	1.96	Heptacosane
		31.721	1.72	Eicosane, 10-Heptyl-10-octyl
	15	15.480	10.13	Hexacosane
		17.889	2.33	Hexadecane, 1-Chloro-acetamide
		18.192	6.60	Tetracosane
		18.489	24.66	8-Heptadecene
		22.143	17.35	9-Octadecenoic acid, Methyl ester
		27.016	14.30	[1,2′-Binaphthalene]-5,5′,8,8′-tetrone, Tetracosane, Pentacosane

Table 4. Major TPH in soil samples A–E on the 5th and 15th days.

4. Conclusions

There was noticeable reduction in the amounts of Zn, Pb, Ni, and Cu in the soil samples. Sorghum husk treatment was the most effective in reducing the concentrations of the heavy metals as there was total removal of lead using sorghum husk and 97.5, 96.1, and 99.3% removal of Zn, Ni, and Cu, respectively. There was also relatively high reduction in heavy metals in the control soil sample. Total petroleum hydrocarbon content of soil was analyzed using GC-MS, 10 and 15 days following treatment. There was reduction in the amount of TPH present after 10 days, further reduction was observed after 15 days. The emergence of various oxidized hydrocarbon compounds such as esters, aldehydes, and carboxylic acids is proof of biodegradation of petroleum hydrocarbons by microorganisms present in the poultry manure and sorghum husk. There was also the presence of amine groups in small amounts, which could be from the addition of the poultry manure. The poultry manure and sorghum husk treatment (1:1) seem to be the most effective in reducing soil TPH. The carbon chain length of the compounds in this group was shorter and majority of them were oxidized. Control group A also exhibited high TPH removal indicating that the provision of a suitable environment for microbial growth is just enough to initiate rapid bioremediation.

Author details

Feyisayo V. Adams*, Maryam F. Awode and Bolade O. Agboola

*Address all correspondence to: feyikayo@gmail.com

Department of Petroleum Chemistry, American University of Nigeria, Yola, Nigeria

References

[1] Hatzikioseyian A. Principles of bioremediation processes. In: Płaza G, editor. Trends in Bioremediation and Phytoremediation. India: Research Signpost; 2010. 23 p. ISBN: 978-81-308-0424-8. Available from: http://old-2017.metal.ntua.gr/uploads/2746/941/Plaza3_prepub.pdf

[2] Adelana SO, Adeosun TA, Adesina AO, Ojuroye MO. Environmental pollution and remediation: challenges and management of oil spillage in the Nigerian coastal areas. American Journal of Scientific and Industrial Research. 2011;**2**:834-845. DOI: 10.5251/ajsir.2011.2.6.834.845

[3] Joutey NT, Bahafid W, Sayel H, El Ghachtouli N. Biodegradation: Involved microorganisms and genetically engineered microorganisms. Rijeka: Intech; 2013. pp. 289-320. DOI: 10.5772/56194

[4] Adams FV, Niyomugabo A, Sylvester OP. Bioremediation of crude oil contaminated soil using agricultural wastes. Procedia Manufacturing. 2017;**7**:459-464. DOI: doi.org/10.1016/j.promfg.2016.12.037

[5] Hidayah WR, Mangkoedihardjo S. Rice husk for bioremediation of chromium(VI) polluted soil. International Journal of Academic Research. 2010;**2**:35-38. DOI: dx.doi.org/10.7813/2075-4124.2010/2-6/A.5

[6] Abioye PO, Alonge OA, Joshua U, Ijah J. Biodegradation of crude oil in soil amended with melon shell. AU Journal of Technology. 2009;**13**:34-38. DOI: doi.org/10.1155/2012/587041

[7] Xu Y, Lu M. Bioremediation of crude oil-contaminated soil: Comparison of different biostimulation and bioaugmentation treatments. Journal of Hazardous Materials. 2010; **183**:395-401. DOI: 10.1016/j.jhazmat.2010.07.038

[8] Ogboghodo IA, Erebor EB, Osemwota IO, Isitekhale HH. The effects of application of poultry manure to crude oil polluted soils on maize (zea mays) growth and soil properties. Environmental Monitoring and Assessment. 2004;**96**:153-161. DOI: 10.1023/B:EMAS.0000031724.22352.af

[9] Okafor UC, Orji MU, Nwankwegu AS, Anaukwu CG, Onuorah SC, Archibong EJ, Obika IE, Agu KC. Effect of chicken droppings amendment on bioremediation of crude oil polluted. European Journal of Experimental Biology. 2016;**6**:62-68

[10] Schumacher BA. Methods for the Determination of Total Organic Carbon (toc) in Soils and Sediments. Washington, DC: U.S. Environmental Protection Agency; EPA/600/R-02/069 (NTIS PB2003-100822). 2002. Available from: https://cfpub.epa.gov/si/si_public_record_report.cfm?dirEntryId=63371

[11] Department of Sustainable Natural Resources. Organic Carbon. In Soil Survey Standard Test Method [Internet]. Available from: https://www.researchgate.net/file.PostFileLoader.html?id=58bd486596b7e4d5a05b2916&assetKey=AS%3A468874685620225%401488799844922 [Accessed: Apr 18, 2017]

[12] Jain PK, Gupta VK, Gaur RK, Lowry M, Jaroli DP, Chauhan UK. Bioremediation of petroleum oil contaminated soil and water. Research Journal of Environmental Toxicology. 2011;**5**:1-26. DOI: 10.3923/rjet.2011.1.26

[13] Olajire AA, Essien JP. Aerobic degradation of petroleum components by microbial consortia. Journal of Petroleum & Environmental Biotechnology. 2014;**5**:1-22. DOI: 10.4172/2157-7463.1000195

[14] Maier RM, Pepper IL. Earth environments. In: Maier RM, Pepper IL, Gerba CP, editors. Environmental Microbiology. 2nd ed. Elsevier: Academic Press; 2009. 57 p

[15] Ludwig F, Fernandes DM, Mota PRD, Bôas RLV. Electrical conductivity and pH of the substrate solution in gerbera cultivars under fertigation. Horticultura Brasileira. Vitoria da Conquista. 2013;**31**:356-360. DOI: dx.doi.org/10.1590/S0102-05362013000300003

[16] Shim J, Shea PJ, Oh B-T. Stabilization of heavy metals in mining site soil with silica extracted from corn cob. Water, Air, and Soil Pollution. 2014;**225**:2152. DOI: doi.org/10.1007/s11270-014-2152-1

[17] Tatah VS, Otitoju O, Ezeonu CS, Onwurah INE, Ibrahim KLC. Characterization and adsorption isotherm studies of Cd(II) and Pb(II) ions bioremediation from aqueous solution using unmodified sorghum husk. Journal of Applied Biotechnology & Bioengineering. 2017;**2**:1-9. DOI: 10.15406/jabb.2017.02.00034

[18] Imaga C, Abia AA, Igwe JC. Removal of Ni(II), Cu(II) and Zn(II) ions from synthetic wastewater using sorghum hulls as adsorbents. Der Chemica Sinica. 2014;**5**:67-75

[19] Barathi S, Vasudevan N. Utilization of petroleum hydrocarbons by Pseudomonas fluorescence isolated from a petroleum-contaminated soil. Environment International. 2011; **26**:413-416. DOI: doi.org/10.1016/S0160-4120(01)00021-6

[20] Mishra S, Jyot J, Kuhad RC, Lal B. Evaluation of inoculum addition to stimulate in situ bioremediation of oily-sludge-contaminated soil. Applied and Environmental Microbiology. 2001;**67**:1675-1681. DOI: 10.1128/AEM.67.4.1675-1681.2001

[21] Foght JM, Westlake D, Johnson WM, Ridgway HF. Environmental gasoline-utilizing isolates and clinical isolates of Pseudomonas aeruginosa are taxonomically indistinguishable by chemotaxonomic and molecular techniques. Microbiology. 1996;**42**:2333-2340. DOI: 10.1099/00221287-142-9-2333

[22] Bossert I, Bartha R. The fate of petroleum in soil ecosystems. In: Atlas RM, editor. Petroleum Microbiology. New York: Macmillan; 1984. pp. 435-473

[23] Cooney JJ. The fate of petroleum pollutants in freshwater ecosystems. In: Atlas RM, editor. Petroleum Microbiology. New York: MacMillan Publishing Co.; 1984. 399 p

[24] Choi S-C, Kae KK, Jae HS, Sang-Jin K. Evaluation of fertilizer additions to stimulate oil biodegradation in sand seashore mescocosms. Journal of Microbiology and Biotechnology. 2002;**12**:431-436

[25] Kim S, Choi DH, Sim DS, Oh Y. Evaluation of bioremediation effectiveness on crude oil-contaminated sand. Chemosphere. 2005;**59**:845-852. DOI: doi.org/10.1016/j.chemosphere.2004.10.058

[26] Pelletier E, Delille D, Delille B. Crude oil bioremediation in sub-Antarctic intertidal sediments: Chemistry and toxicity of oiled residues. Marine Environmental Research. 2004;**57**:311-327. DOI: doi.org/10.1016/j.marenvres.2003.07.001

[27] Okolo JC, Amadi EN, Odu CTI. Effects of soil treatments containing poultry manure on crude oil degradation in a sandy loam soil. Applied Ecology and Environmental Research. 2005;**3**:47-53

[28] Perry JJ. Microbial metabolism of cyclic alkanes. In: Atlas RM, editor. Petroleum Microbiology. New York: Macmillan Publishing; 1984. pp. 61-98

[29] Verstraete W, Vanloocke R, DeBorger R, Verlindne A. Modeling of the breakdown and the mobilization of hydrocarbons in unsaturated soil layers. In: Sharpley JM, Kalpan AM, editors. Proceedings of 3rd International Biodegradation Symposium. London: Applied Science Publishers; 1975. 98 p

[30] Dibble JT, Bartha R. The effect of iron on the biodegradation of petroleum in seawater. Applied and Environmental Microbiology. 1976;**31**:544-550. Available from: https://www.ncbi.nlm.nih.gov/pmc/articles/PMC169817/pdf/aem00004-0102.pdf

Chapter 7

Heavy Metal Removal with Phytoremediation

Sevinç Adiloğlu

Additional information is available at the end of the chapter

http://dx.doi.org/10.5772/intechopen.70330

Abstract

Heavy metal pollution in agricultural soil is one of the most important environmental problem for the different scientists, recently years. Heavy metal contamination in the agricultural soil is not only pollution but it also has dangerous effect on wild life and human life. The solution of this pollution problem by using classical traditional physical and chemical methods is too expensive. But, phytoremediation method is using for removal of heavy metal from agricultural soils, recently. This method is cheaper than classical traditional physical and chemical methods. Kinds of phytoremediation method are phytoextraction, phytodegradation, phytostabilization, phytovolatilization, rhizodegradation, rhizofiltration, phytohydraulic control, vegetative cover systems, buffer strips and riparian corridors. These kinds of phytoremediation methods were evaluated in this study.

Keywords: phytoremediation, heavy metal, soil, pollution, plant

1. Introduction

Phytoremediation, which is derived from the words "phyto" (plant) and "remediation" (recovery) and has become a term in 1991, can be also defined as "bioremediation," "botanical remediation" and "green remediation." Phytoremediation is a term which is related to ecological remediation technologies that use plants as the main source. With this technology, organic and inorganic substances are removed from the contaminated area by using plants. The effects of this method can be observed in low polluted areas in a short time. The negative aspect is that in heavy contaminated areas the plants cannot be useful in a period of short time [1, 2].

The plants which are identified as metal hyper-accumulators and wild, are able to remove contaminant elements 10–500 times higher compared to the ones that are cultivated [3].

Phytoremediation method is ecological, does not need special equipment during application and provides a re-usable land. The root depths and climatic conditions play an important role in the efficiency of the system. First of all, the soil must be appropriate to the needs of the plant for the removal of the contaminants from the soil by the plant. The pH of the soil is one of the most important parameters. The pH levels of the area must be between 5.8 and 6.5 for the nutrient elements to be taken [4]. The absorption of the contaminants and their accumulation by the plants is presented in the **Figure 1**.

The nutrient element absorption is completed in three stages: (1) the transportation of the nutrients to the root circle and root surface; (2) the absorption of the nutrient ions into the roots; and (3) the transportation of the nutrient ions which entered into the root to the necessary parts by the transmission branches. There are two basic theories in the transportation of the nutrients to the root surface: "Intersection and Contact Change" and "Carbonic Acid Theory" [6].

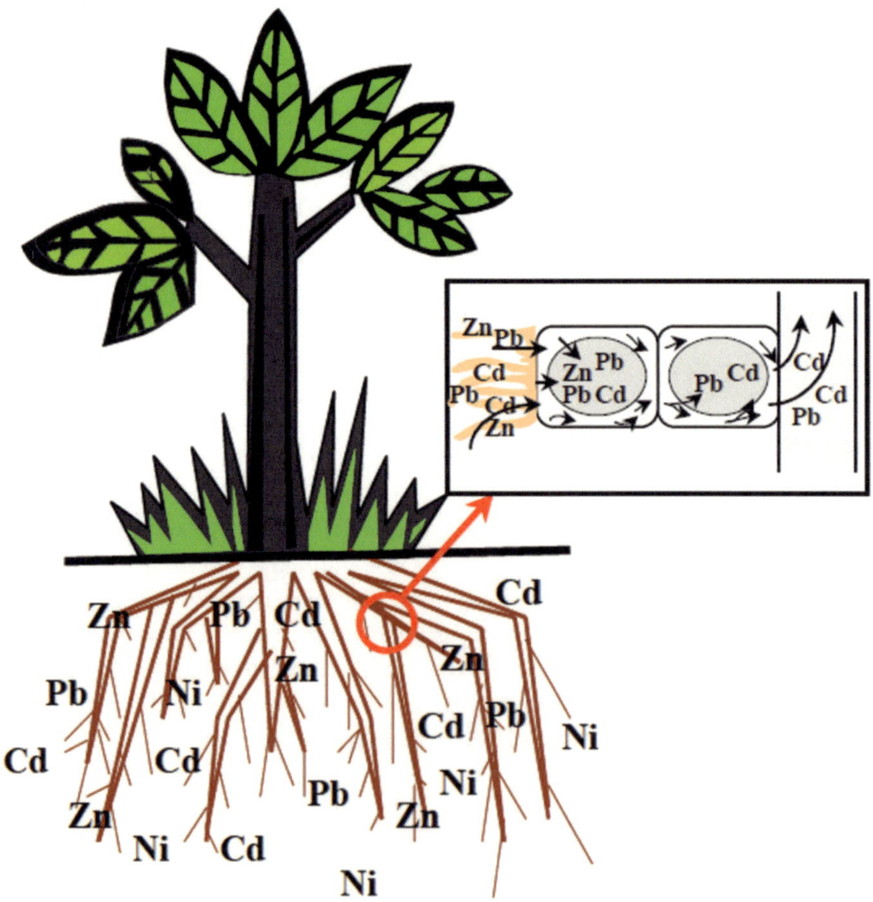

Figure 1. Heavy metal absorption and accumulation in root and shoot of the plant [5].

http://dx.doi.org/10.5772/intechopen.70330

2. Phytoremediation techniques

Phytoremediation, which has become more common in the last 10 years, is a passive technology which is related to soil recovery. Phytoremediation is the use of green plants in the removal of the contaminants from the area or in their recovery [7].

According to Salt et al. [8] phytoremediation techniques can be subcategorized as phytoextraction, phytodegradation, rhizofiltration, phytostabilization, phytovolatilization and rhizodegradation. Phytoremediation techniques are very effective in the sterilization of the areas that are medium-contaminated and have slight risk.

2.1. Phytoextraction (vegetal assimilation)

This technique is used in the absorption of the organic and inorganic contaminants by the roots and the sprigs of the plant. It is a valid method for the recovery of the contaminated areas, in which the plants that are able to absorb metals are chosen and the contaminants are removed from the soil with the harvesting or removal of the plant.

Because this technology's application takes more time compared to other techniques, its application on heavy polluted areas is very hard. Also, a plant which grows in that ecosystem should be chosen. It should not be seasonal, because they will be harvested later. After they are harvested, they are burned in incinerator or exposed to another method with composition [9]. This method which is called phytomining, provides the opportunity for obtaining the mineral ores whose cultivation process is not economic. With this method, gold and nickel are re-gained in the USA [1, 10].

When the plants used in this method are compared to other plants, it can be observed that they can accumulate contaminant elements 100 times more. In this method Brassicacea, Euphorbiacea, Asteraceae, Lamiaceae and Scrophulariaceae and 400 other types are identified which can accumulate heavy metals. The residues of the harvested plants can be isolated by drying, burning, composting and recycling to biological metal minerals [11].

2.2. Phytodegradation (vegetal degradation)

Phytodegradation is a method in which organic contaminants are degraded by the compounds that are produced by plants through metabolic processes. Vegetal degradation can be applied to soil, clay, sediment and underground waters. The most advantageous aspect of the method is that the reduction and degradation occur inside the plant as a physiological process, and do not depend on microorganisms, while the emergence of toxic intermediate and end-use products, and the difficulty of their detection create a disadvantage [10].

The absorption of the organic compounds into the plant depends on the plant type, the residence duration of the contaminant element in the soil, and the soil's physical and chemical form. The easily dissolved compounds are difficult to absorb. Plant enzymes are known to be able to degrade hazardous substances such as herbicides, munitions wastes and chlorinated solvents (trichloroethane (TCE)) [11].

2.3. Phytostabilization (root stabilization)

This method is used in the stabilization of soil. Phytostabilization plants are able to tolerate heavy metal levels and immobilize the metals through sorption, sedimentation, complexation or reduction of metal valences. The contamination factors in soil occur as a result of the immobilization of the contaminants around the plant roots, their accumulation by the roots, cohesion or sedimentation around the roots [12].

Wang et al. [13] conducted a research on the development and Cu absorption of corn plant (*Zea mays* L.) which is inoculated or non-inoculated by *Acaulospora mellea*, an arbuscular mycorrhizal fungus, by using different doses of Cu-applied pots in laboratory conditions. They concluded that the low absorption of the plants in the high concentrated Cu pots results from the soil's pH value. They observed that the concentration and the structures of the organic acids in the soil such as malic acid, citric acid and oxalate acid were modified by the fungus. The researchers revealed that *Acaulospora mellea* is not suitable for the phytoextraction of copper by the corn plant; however, mycorrhizal plants are more applicable for phytoextraction because of their high capacity of Cu absorption in their roots.

On the other hand, contaminants' transportation by wind, water erosion, washing out or soil dissemination can be prevented. In a system which is closely related to the plant's root environment microbiology and chemistry, the plant is able to modify the contaminant factor's form into non-resoluble or non-transported in water [1, 2].

2.4. Phytovolatilization (vegetal evaporation)

The root depth is very crucial in phytovolatilization. If it is about underground waters, the roots should be deep. To sterilize contaminated underground waters, the water can also be pumped to the ground to provide absorption for the surface roots. The most important aspect of this method is the transformation of the excessive toxic compounds (mercury contained compounds) into less toxic forms. However, the potential release of these hazardous and toxic materials into the atmosphere is a disadvantage [1]. The contaminants can be removed from the plant by transpiration or evaporation. As a well-known fact, water is carried from the roots to the leaves with the help of vascular system; therefore, the contaminants are released to the air through evaporation or volatilization. Poplar tree can be an example for this mechanism [9].

Ghosh and Singh [14] pointed out that some plants such as *Brassica juncea* and *Arabidopsis thaliana* can release heavy metals to the atmosphere with phytovolatilization by absorbing and transforming them into gas form.

Some types of trees such as *Populus* and *Salix* are often used in phytovolatilization because of their capacity to take contaminants with phytoremediation [15].

2.5. Rhizodegradation (the use of roots for degradation)

Rhizodegradation is the decomposition of the organic contaminants in soil surrounding the roots of the plants as a result of microorganism activities. There are amino acids, sugar, organic acid, sterol, fat acids, growing factors, nucleotide, flavanone and enzymes which are released from the plant's roots and affect the microbial activities in the surrounding area of the roots. The most important benefit of Rhizodegradation method is the dissolution of the contaminants in their natural environment [1, 2].

Pesticides (herbicide, insecticide), benzene, toluene, ethylbenzene, xylene (BTEX), total petroleum hydrocarbon (TPH), polycyclic aromatic hydrocarbons (PAH), surface active substances, chlorinated solvents (TCE, TCA), pentachlorophenol (PCP) polychlorinated biphenyls (PCB) can be exemplified as contaminants that can be dissolved with Rhizodegradation. Mint (*Mentha spicata* L.), red berry (*Morus rubra* L.), lucerne (*Medicago sativa* L.), and reedmace (*Typha latifolia* L.) are used in Rhizodegradation method [1, 4, 16, 17].

2.6. Rhizofiltration (the use of roots for filtration)

In rhizofiltration method, contaminants cling to the roots or absorbed by the roots in accordance with biotic and abiotic processes. During these processes, contaminants may be taken or transported by the plant. What is important is to maintain the immobilization of the contaminants in or on the plants. Later on, the contaminants can be taken from the plants with different methods. This method is applied to underground waters, surface waters and waste waters [4, 18].

Rhizofiltration is used to remove the radioactive substances or metals from the contaminated waters. The plants which are used in this method are directly planted on the contaminated soil and the contaminant's adaptation is ensured. The plants are raised hydroponically in clean water instead of soil until they have a wide root system. The rooted plants are transported to the contaminated water source in order to make them adapt to their new environment. When the roots become saturated the plants are harvested. This method provides an opportunity for the use of terrestrial and aquatic plants. It is also used in basins, tanks, and ponds besides natural environment [8, 12].

2.7. Phytohydraulic control

Hydraulic control is a method which prevents and controls the accumulation and transportation of the contaminants by using plants. This method is applied to both underground and surface waters. The advantageous aspect of this method is the wide impact area because of the expansion of the roots without any artificial system. However, the instability of water absorption depending on season and climate is a disadvantage.

According to Pivetz [10], a 5-year-old *Populus* tree can absorb 100–200 liters of water in a day. A single salix tree's amount of perspiration is claimed to be 5000 gallon of water in a day. Salix, hybrid populous and Eucalyptus can be used in this method. The Phytohydraulic control method is generally used in the dissolution of organic and inorganic water-soluble contaminants [1].

2.8. Vegetative cover systems

Vegetative cover is a method in which the contaminants are controlled by the long-term and self-growing vegetative system. Vegetative cover systems expand over or inside of the substances with environmental risks and require minimum care. Vegetative cover is generally set up as barriers that prevent the expansion of contamination [1].

Vegetative covers are applied to the contaminant area or to the surface around the units that spread the contaminants. This system is planned to be used in the USA as an alternative in covering solid waste storage areas considering it as a cheap ecosystem which minimizes surface erosion by regenerating itself [1, 4, 10]. However, the constant control of the necessary long-term maintenance of the cover system creates a disadvantage, because some plant types may dominate the others in the course of time [19].

2.9. Buffer strips and riparian corridors

Buffer strips and riparian corridors are systems in which the suitable plants are planted in stripes throughout stream rank in order to remove the contaminants in underground and surface waters that stream towards the rivers [4, 10].

This method prevents the water contaminants to spread and interfuse into ground water. The studies conducted in Canada revealed that this system removes soil erosion up to 90%, and herbicides up to 42–70% [20].Buffer strips method is mostly used in many countries in the removal of contamination caused by fertilizers and pesticides. *Populus* is the most used tree in this method [1].

This method is mainly used in the removal the contaminants in underground and surface waters that stream towards the rivers by plantation in stripes throughout stream rank. Therefore, the water contaminants are prevented to spread and interfuse into ground water. On the other hand, erosion is controlled and sediment is reduced. The studies have shown that this system removes sediment in water up to 71–90%, nitrogen up to 67–96%, phosphorus up to 27–97% pesticides up to 8–100% and fecal coliforms up to 70–74% [19, 20].

3. Phytoextraction of heavy metals with Canola in model field

We performed the phytoextraction with canola in the model fields to estimate ant effectiveness.

3.1. Experimental conditions

The experiment is conducted on the research fields of Faculty of Agriculture, Namık Kemal University, Turkey, according to the randomized block design with 3 replicates. 100 mg/kg Co, Cr, Ni and Pb ions are taken from $CoSO_4$, $Cr(NO_3)_3$, $NiSO_4$ and $Pb(NO_3)_2$ compounds and applied to the soil as contaminants. There are 51 parcels with 4 contaminants ($CoSO_4$, $Cr(NO_3)_3$, $NiSO_4$ and $Pb(NO_3)_2$) × 4 chelate doses (EDTA) (0, 5, 10 and 15 mmol/kg) × 3 replicates + 3 control. The control parcels in which the contaminants and chelate are not applied are organized as three replicates. Each parcel in the test are sized as (3 × 1.2 m): 3.6 m^2 including four rows (for instance for lead element Pb_{EDTA0}, Pb_{EDTA5}, Pb_{EDTA10} and Pb_{EDTA15}). The row distance in each parcel is 30 cm, the distance between each parcel is 0.5 m and between each block is 1.5 m. The height of each block is 3 m and the width is 31.2 m (there are 4 parcels in each block). Therefore, a whole block is of 93.6 m^2 and the total test area is of 284.6 m^2, including the distances between the blocks (**Figure 2**).

Figure 2. Different views from experiment fields (original).

3.2. Physical and chemical characteristics of the test soils

The physical and chemical characteristics of the samples taken from the testing field are presented in **Table 1**. According to the table, the soil's pH is neutral, low lime and has insufficient organic matter. Its available phosphorus content is sufficient as well as the exchangeable potassium. The amount of available iron is average, available copper and manganese is sufficient, and available zinc is insufficient. Also, the testing soil is classified as clay in terms of texture [21, 22].

Soil properties	Unit	Values
pH (soil: water=1:2.5)		6.81
EC (x10^6)	dS/m	128.3
$CaCO_3$	%	2.40
Organic matter	%	1.88
P_2O_5	kg/da	11.42
K_2O	kg/da	25.32
Fe	mg/kg	3.46
Cu	mg/kg	0.63
Zn	mg/kg	0.40
Mn	mg/kg	5.72
Clay	%	42.98
Silt	%	25.44
Sand	%	31.58
Texture class		C

Table 1. Some physical and chemical characteristics of the testing soil [23].

3.3. The amount of heavy metal (Cr, Co, Ni and Pb) before the test and after the incubation

The extractable Cr, Co, Ni and Pb contents of the testing fields are identified before the heavy metals are applied to the soil and they are presented in **Table 2**. 100 mg/kg Cr, Co, Ni and Pb are applied to the testing field and left to incubation for a month. The extractable heavy metal contents are determined after the incubation. T-test is applied to the results and the standard error values are given in **Table 2**.

When the **Table 2** is examined, it can be observed that the amount of extractable heavy metal contents before the test is acceptable and does not have any contaminant characteristic [1]. A remarkable increase is observed as a result of 100 mg/kg Cr, Co, Ni, Pb heavy metal application upon one-month incubation. These increases are determined as 1% significant statistically.

3.4. The effects of EDTA applications on heavy metal contents (Cr, Co, Ni, Pb) of the root and shoot of the plant

The effects of increasing doses of EDTA applications on heavy metal contents (Cr, Co, Ni, Pb) of the root and shoot of the plant are presented in **Table 3**.

The amount of Cr in the roots and shoot of the canola plant which grows on Cr-applied fields rapidly increased after 0 mmol/kg EDTA dose and reached the highest level with 15 mmol/kg EDTA dose. These increases were determined as 1% significant statistically.

Similar to the Chrome element, the amount of cobalt in the root and shoot of canola plant which grows on the cobalt-polluted field increased with EDTA applications. These increases were determined as 1% significant statistically (**Table 3**).

The amount of nickel in canola plant which grows on the field that has been polluted with nickel increased with EDTA applications, and the highest levels were achieved on the parcels which were applied with 15 mmol/kg EDTA dose. It can be concluded that with the increasing doses of EDTA, the Ni concentration in the roots and shoot of the plant has also increased. These increases were determined as 1% significant statistically and various groups are formed in Duncan multiple comparison test (**Table 3**).

The amount of Pb in canola plant which grows on the field that has been polluted with Pb increased with EDTA applications, and the highest levels were achieved on the parcels which

Heavy metal	Before polluting	After polluting
Cr	$0.10 \pm 0.01^{**}$	$5.80 \pm 0.07^{**}$
Co	$0.08 \pm 0.05^{**}$	$2.25 \pm 0.02^{**}$
Ni	$0.95 \pm 0.05^{**}$	$6.20 \pm 0.13^{**}$
Pb	$0.93 \pm 0.01^{**}$	$7.52 \pm 0.04^{**}$

** $p< 0.01$.

Table 2. The extractable heavy metal contents before and after the application of the contaminants to the testing soil.

Canola				
	Chrome (Cr)		**Cobalt (Co)**	
EDTA application	**Root**	**Shoot**	**Root**	**Shoot**
Control	2.75 ± 0.57a	4.23 ± 0.58a	1.95 ± 0.02a	2.18 ± 0.04a
0 mmol/kg	14.70 ± 007b	8.25 ± 0.56a	13.30 ± 0.08b	11.56 ± 1.39b
5 mmol/kg	29.16 ± 0.58c	26.42 ± 0.21b	28.60 ± 1.50c	24.45 ± 0.59c
10 mmol/kg	55.12 ± 0.99d	40.45 ± 1.05c	51.40 ± 2.61d	39.12 ± 1.14d
15 mmol/kg	70.50 ± 1.92e	52.20 ± 1.50d	75.40 ± 1.56e	45.20 ± 0.60e
	Nickel (Ni)		**Lead (Pb)**	
EDTA application	**Root**	**Shoot**	**Root**	**Shoot**
Control	4.76 ± 0.11a	5.83 ± 0.63a	3.78 ± 0.45a	4.96 ± 0.56a
0 mmol/kg	24.43 ± 0.13b	13.12 ± 0.97b	24.50 ± 0.97b	18.90 ± 0.54b
5 mmol/kg	49.65 ± 0.34c	37.60 ± 0.56c	41.40 ± 0.66c	35.20 ± 0.05c
10 mmol/kg	77.80 ± 0.60d	61.40 ± 0.28d	87.80 ± 0.90d	63.14 ± 1.14d
15 mmol/kg	85.30 ± 1.01e	65.10 ± 0.057e	95.40 ± 0.17e	70.12 ± 0.01e

*Each heavy metal element, root and shoot is examined separately with three replicates.

Table 3. The effects of EDTA applications on the amount of heavy metals (Cr, Co, Ni and Pb) in the roots and shoot of canola plant that were grew* (mg/kg) [23–26].

were applied with 15 mmol/kg EDTA dose. It can be concluded that with the increasing doses of EDTA, the Pb concentration in the roots and shoot of the plant has also increased. These increases were determined as 1% significant statistically and various groups are formed in Duncan multiple comparison test (**Table 3**).

In **Table 3**, the amount of heavy metals (Cr, Co, Ni and Pb) in canola plant's roots and shoot, Duncan multiple comparison test and standard error values are presented. According to **Table 3**, the amount of heavy metals in the roots is higher than the heavy metals in shoot.

According to **Table 3**, on the fields which are applied with 0 mmol/kg EDTA doses, the amount of Cr, Co, Ni and Pb heavy metals in the roots and shoot of the canola plant is lowest and the highest heavy metal level can be detected in 15 mmol/kg EDTA dose. The results are equal to other research results on this subject. The researchers have explained that the solubility and absorption of the heavy metals get easier for the plant with the increasing doses of EDTA application [27–31].

According to the results of this research, EDTA applications should be conducted by growing hyper-accumulator plants on the soils in order to decrease Cr, Co, Ni and Pb heavy metal contamination under the toxicity levels. Because, it has been proved that with the increasing doses of EDTA application some heavy metals (Cr, Co, Ni and Pb) can be removed from soil with phytoremediation (Phytoextraction) method.

4. Conclusion

According to the field experiment results, the amount of Cr, Co, Ni and Pb heavy metals, removed from the soil by the canola plant, increased with the increasing of EDTA applications. This increases and decreases were found statistically significant at the level of 1%. It was an expected rate, because application of chelates like EDTA on soil accelerates solubility of some heavy metals (Cr, Co, Ni and Pb) in soils and their absorption by plants. It was relatively expensive and limited in use to remove some heavy metals like Cr, Co, Ni and Pb which cause pollution in agricultural lands and are particularly results of industry-related human activities, from the soil by classical physiochemical methods.

Therefore, it becomes gradually important to remove Cr, Co, Ni and Pb heavy metals naturally from the agricultural lands by increasing its mobility in soil, with the help of various hyper accumulator plants such as canola, which are relatively inexpensive and practical in use but changeable concerning the concentrations and types of heavy metals.

Author details

Sevinç Adiloğlu

Address all correspondence to: sadiloglu@hotmail.com

Department of Soil Science and Plant Nutrition, Faculty of Agriculture, Namık Kemal University, Tekirdağ, Turkey

References

[1] EPA, (Environmental Protection Agency). Introduction to Phytoremediation, EPA/600/R-99/107. Ohio 45268, USA: National Risk Management Research Laboratory Office of Research and Development U.S. Environmental Protection Agency Cincinnati; 2000

[2] Yıldız N. Principles of Plant Nutrition and Disorders of Plant Nutrition in Plants. Erzurum: Atatürk University Agricultural Faculty. Eser Offset Printing; 2008. p. 304

[3] Ow DW. Heavy metal tolerance genes: Prospective tools for bioremediation resources. Conservation Recycling. 1996;**18**:135-149

[4] Vanlı Ö. Pb, Cd, B Elementlerinin Topraklardan Şelat Destekli Fitoremediasyon Yöntemiyle Giderilmesi. ABD: İTÜ, Fen Bil. Enst. Çevre Müh, Yüksek Lisans Tezi; 2007

[5] Eke M. Nikel Hiperakümülatörü *Thlaspi Elegans* Boiss'den Nikelin Asitle Ekstraksiyonu ve Elektrokimyasal Yolla Metal Olarak Geri Kazanımının Araştırılması. Mersin: Mersin Üniversitesi Fen Bilimleri Enstitüsü Çevre Anabilim Dalı Yüksek Lisans Tezi; 2010

[6] Karaman MR, Adiloğlu A, Brohi R, Güneş A, İnal A, Kaplan M, Katkat V, Korkmaz A, Okur N, Ortaş İ, Saltalı K, Taban S, Turan M, Tüfenkçi Ş, Eraslan F. ve Zengin M. Bitki Besleme. ISBN 978-605-87103-2-0 Dumat Ofset. Şti., Ankara: Matbacılık San. Tic. Ltd; 2012

[7] Raskin I, Smith RD, Salt DE. Phytoremediation of metals using plants to remove pollutants from the environment. Current Opinion in Biotechnology. 1997;**8**:221-226

[8] Salt DE, Smith RD, Raskin L. Phytoextraction of cadmium with *Thlaspi caerulescens*. Plant Molecular Biology. 1998;**49**:643-668

[9] EPA. Contaminants and Remedial Options at Select Metals – Contaminated Sites, EPA/540/R-95/512. National Risk Management Research Laboratory Office of Research and Development U.S. Environmental Protection Agency. Work EPA, (Environmental Protection Agency), 2000. Introduction to Phytoremediation, EPA/600/R-99/107, National Risk Management Research Laboratory Office of Research and Development U.S. Environmental Protection Agency Cincinnati, Ohio 45268, USA; 1995

[10] Pivetz BE. Ground Water Issue: Phytoremediation of Contaminated Soil and Ground Water at Hazardous Waste Sites. United States Environmental Protection Agency, EPA, 540/S–01/500; 2001. p. 36

[11] Memon AR, Aktopraklıgil D, Özdemir A, Vertii A. Heavy Metal Accumulation and Detoxification Mechanisms in Plants. TÜBITAK MAM. Kocaeli, Turkey: Institute for genetic Engineering and Biotechnology; 2000

[12] Mirsal IA. Soil pollution: Origin, Monitoring and Remediation. Springer–Verlag Berlin Heidelberg, Germany; 2004

[13] Türkoğlu B. Toprak Kirlenmesi Ve Kirlenmiş Toprakların Islahı. Adana: Çukurova Üniversitesi Fen Bilimleri Üniversitesi Toprak Anabilim Dalı Yüksek Lisan Tezi; 2006

[14] Wang FY, Lın XG, Yın R. Inoculation with arbuscular mycorrhizal fungus acaulospora mellea decreases Cu phytoextraction by maize from Cu–contaminated soil. Pedobiologia. 2007;**51**:99-109

[15] Ghosh M, Singh SP. A review on phytoremediation of heavy metals and utilization of its byproducts. Applied Ecology and Environmental Research. 2005;**3**:1-18

[16] Pulford ID, Watson C. Phytoremediation of heavy metal-contaminated land by trees: A review. Environment International. 2003;**29**:529-540

[17] Aybar M, Bilgin A, Sağlam B. Removing Heavy Metals from the Soil with Phytoremediation. Artvin Çoruh University Journal of Natural Hazards and Environment. 2015;**1**:59-65

[18] Söğüt Z, Zaimoğlu Z, Erdoğan RK, ve Doğan S. Su kalitesinin arttırılmasında bitki kullanımı (yeşil ıslah-Phytoremediation). Türkiye'nin Kıyı ve Deniz alanları IV. Ulusal Konferansı 5-8 Kasım 2002. Dokuz Eylül Üniversitesi, İzmir. Bildiriler Kitabı. II. Cilt: 2002. 1007-1016

[19] Kalkan H, Oraman Ş, ve Kaplan M. Kirlenmiş Arazilerin Islah Edilmesinde Fitoremidasyon Tekniği. Selçuk Üniversitesi Selçuk Tarım ve Gıda Bilimleri Dergis.,i. 2011; **25**(4):103-108

[20] Gabor TS, North AK, Ross LCM, Murkin HR, Anderson JS, Turner MA. Beyond the Pipe: The Importance of Wetlands and Upland Conservation Practises in Watershed Management: Function and Values for Water Quality and Quantity. Ducks Unlimited Canada; 2001. p. 55

[21] Alparslan M, Güneş A, ve İnal A. Deneme Tekniği. Ankara: A.Ü. Ziraat Fakültesi Yayın No:1501; 2005

[22] FAO. Micronutrient, Assesment and the Country Level: An International Study. Rome, Italy: FAO Soils Bulletin 63; 1990

[23] Adiloğlu S, Sağlam MS. Removal of lead (Pb) pollution in agricultural areas by phytoremediation method with the use of Canola (*Brassica napus* L.) plant. Proceedings of the 13th International Conference of Environmental Science and Technology, Athens, Greece; 2013

[24] Adiloğlu S, Sağlam MT, Süme A. Chrome (Cr) pollution in agricultural areas improvement by phytoremediation method with Canola (*Brassica napus* L.) plant growing. Journal of Essential Oil Bearing Plants. 2015;**18**(5):1180-1186

[25] Adiloğlu S, Sağlam MT, Adiloğlu A, Sümer A. Removal of nickel (Ni) from agricultural field soils by phytoremediation using Canola (*Brassica napus* L.). Desalination and Water Treatment. 2016;**57**(6):2383-2388

[26] Adiloğlu S. Using phytoremediation with Canola to remove cobalt from agricultural soils. Polish Journal of Environmental Studies. 2016;**25**(6):2251-2254

[27] McGrath SP, Shen ZG, Zhao FJ. Heavy metal uptake and chemical changes in the rhizosphere of *Thlaspi caerulescens* and *Thlaspi ochroleucum* grown in contaminated soils. Plant and Soil. 1997;**188**:153-159

[28] Epstein A, Gussman CD, Blaylock MJ, Yermiyahu U, Huang JW, Kapulnik Y, Orser C. EDTA and Pb-EDTA accumulation in *Brassica juncea* grown in Pb-amended soil. Plant and Soil. 1999;**208**:87-94

[29] Kirkham MB. EDTA-facilitated phytoremediation of soil with heavy metals from sewage sludge. International Journal of Phytoremediation. 2000;**2**:159-172

[30] Liphadzi MS, Kirkham MB, Mankin KR, Paulsen GM. EDTA-assisted heavy –metal uptake by poplar and sunflower grown at a long-term sewagesludge farm. Plant and Soil. 2003;**257**:171-182

[31] Esringü A. Toprakta Kirliliğe Sebep Olan Bazı Ağır Metallerin Kanola (*Brassica nopus*) ve Hardal (*Brassica juncea*) Bitkileri İle Uzaklaştırılması. Atatürk Üniversitesi Fen Bilimleri Enst. Toprak Anabilim Dalı, Yüksek Lisans Tezi; 2005

Chapter 8

Potential and Constraints of Macrophyte Manipulation for Shallow Lake Management

Zeljka Rudic, Bojana Vujovic, Ljubinko Jovanovic,
Dragan Kiković, Igor Kljujev, Mile Bozic and
Vera Raicevic

Additional information is available at the end of the chapter

http://dx.doi.org/10.5772/intechopen.74046

Abstract

Palic and Ludas lakes are shallow Pannonian lakes, which have undergone rapid eutrophication. High concentrations of nutrients, along with significantly high values of fecal indicators in water samples, indicate that the Palic-Ludas canal, which connects them, contributes to the pollution of the recipient Lake Ludas, a central part of a special nature reserve. In order to improve water quality in the canal and decrease nutrient load of Lake Ludas, it is suggested to use environmentally friendly solutions, so-called bio-barriers (bio-bridges and biobanks), which will contribute to self-purification efficiency. The given model assumes 10 zones with bio-bridges and 0.4 ha covered by biobanks, using plants common in the area, e.g. *Typha* spp., *Phragmites* spp., *Juncus* spp., *Scirpus* spp. and *Carex* spp. The main disadvantage of this technology is occupation of large area; thus, the solution seems to be undersized. It removes 4% TN/year and 8% TP/year of the total amount necessary to achieve good ecological status. Nevertheless, the role of bio-bridges in permanent nutrient removal, through preventing the deposition of organic matter at the bottom sediment and later return of nutrients in water, together with indirect influence through enhancement of biodiversity should not be underestimated.

Keywords: bio-bridge, bio-barrier, ecoremediation, eutrophication, macrophyte species, lake degradation, Pannonian lakes

1. Introduction

Shallow lakes have an important role for the humanity and maintenance of environmental quality. They contribute to water supply and represent a resource for plant, fish and amphibian biodiversity, and also they are habitat for different bird species and have a recreational and socio-economic significance and esthetic values. However, shallow aquatic ecosystems are affected by numerous stressors as eutrophication, pollution, invasion of different species, drought, uncontrolled fishery and climate changes. Mentioned stressors often have a synergistic impact that enlarges the consequences and contributes to accelerated degradation of shallow lake ecosystem. Ecology of shallow lakes differs in comparison to deep stratified lakes according to many characteristics, so their shift from the clean state with dominated macrophytes to the status with dominated algae happens very fast during eutrophication process [1]. Eutrophication or nutrient enrichment affects the entire aquatic system by altering trophic structure, biodiversity and biogeochemical cycles, as well as seasonal dynamics. Many shallow lakes lost their ecological value in the past decades, through occurrence of eutrophication, acidification, invasion of new species and climate change [2–4]. Today, water pollution caused by excessive input of N and P from agriculture, urbanization and industrial discharge has a very negative effect and becomes a global issue [5]. Eutrophication is the critical problem impairing surface water quality (especially in lakes and reservoirs), and the effective control of lake eutrophication is needed. European member states have an obligation to develop ecological system based on estimation of biological communities defined by a certain number of elements of biological quality and to adjust water management with the aim to achieve at least "good" ecological status for all water bodies by 2015 and ultimately to 2027 [6]. Eutrophication has many undesirable side effects, major economic costs and transnational implications [7, 8]. Large efforts have been made to combat eutrophication by reducing the external loading of phosphorus in many countries of Europe and North America [9]. In the USA, annual costs are about $ 2.2 billion as a result of freshwater eutrophication [10].

Implementation of different measures that are mainly based on control of nutrient input and interventions on complex food web in lakes is necessary to re-establish their function [11]. Reduction of nutrient input as a single measure is not sufficient to reverse the effects of eutrophication in shallow lakes [9, 12]. Among other restoration methods, biomanipulation has been widely studied and applied to control eutrophication because of its high and more important long-term efficiency [13–15]. Several actions are usually necessary to achieve and maintain "clear" state in shallow lakes, as reduction of external loading, internal phosphorous loading and the presence of stable submerged macrophyte communities [16]. Aquatic macrophytes have an important role in cleaning eutrophic runoff water from agriculture and urban areas [17] by employing several mechanisms in aquatic ecosystems, such as competing with phytoplankton for nutrients, reducing the resuspension of the sediment and providing a refuge for zooplankton; therefore, their potential use has been extensively studied [13, 15, 17, 18]. In sync with ecological interactions, many physicochemical and biological processes such as sedimentation, filtration, precipitation, plant uptake and microbial decomposition are frequently involved in ecological engineering [19], such as creating wetlands [20] and wastewater treatment [21] improving ecosystem quality [22].

The scope of this chapter is to highlight the importance of the preservation of natural diversity of shallow lakes and how to maintain satisfactory environmental quality through macrophyte manipulation. Researchers suggest different restoration methods for shallow lakes with disturbed natural balance, using chemical or biological treatments, although there is no unique method for their restoration. Special cases are protected areas and special nature reserves, similarly, where limited actions can be performed.

2. Characteristics of shallow Pannonian lakes

Most of the lakes of the Pannonian Plain in Hungary, eastern Austria and northern Serbia are originally saline with water of carbonate type [23]. However, with urbanization and industrial development, their water chemistry has changed, and even some problems occurred, usually due to pollution. One of the examples is Lake Balaton in Hungary. Significant socio-economic development on the watershed since the 1960s resulted in increasing external loads and decline in water quality [24]. To retain the nutrients from the lake, the solution was to design and implement Kis-Balaton Water Protection System, i.e. to do wetland reconstruction at the lower part of the River Zala, which is the main tributary of the Lake Balaton [25]. As reported by Tatrai et al. [25], the efficiency of nutrient retention of the protection system was higher than expected. In part of Pannonian Plain that lies in Vojvodina (north Serbia), there are 75 shallow alkaline lakes and ponds [26]. They usually lie in depressions, often with depths less than 0.5 m. Two relatively large shallow lakes with exceptional natural value that are affected by human influence, lakes Palic and Ludas, are the subject of this study.

Pannonian lakes, Palic and Ludas, are very shallow lakes located at the periphery of the Suboticko-Horgoska Pescara (sand area), at the north of the Republic of Serbia. Both lakes are exceptional, since they are home to diverse habitats (aquatic, swamp, meadow and steppe) with a number of plant and animal species, some of which are strictly protected. They are a part of IBA (Important Bird and Biodiversity Area) site. Lake Palic is also a part of a natural park, which is a protected area of local significance. Lake Ludas belongs to the first category of protection, as a natural area of exceptional significance, and it is on the list of Ramsar sites.

The surrounding area is characterized by continental climate, with severe winters, hot summers and irregular distribution of precipitation. The average air temperature is 10.8°C, air humidity 69% and air pressure 1007 mbar. The annual number of rainy days is 105, and mean annual precipitation value amounts 561.1 mm, while the number of days with snow cover is 59 per year. This area is also characterized by strong winds (wind speed more than 6 Bf, i.e. 34 km/h), during 104 days per year. The highest precipitation output is during late spring and beginning of summer (June), and the lowest is in winter time—January, February and March (**Figure 1**). Dry season begins in July and lasts until October (**Figure 1**).

2.1. Lakes Palic and Ludas

Lake Palic is a shallow Pannonian lake, created million years ago, during creation of pits and dunes by wind erosion. In the past, the lake was recharged mostly by atmospheric precipitation. It covers surface of 5.65 km^2. During the 1970's, it was split into four sectors to prolong retention time, as a part of former restoration project. The volume of the lake is approximately 11.5×10^6 m^3.

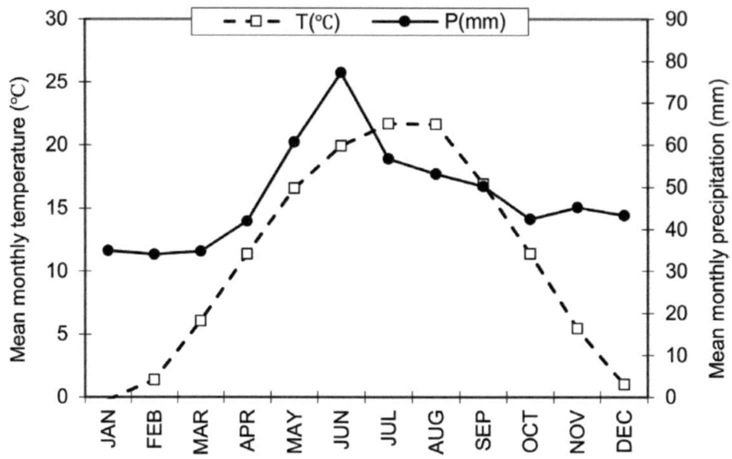

Figure 1. Climate diagram for meteorological station "Palic" (1978–2016).

Lake Ludas is oblong, with wide, low and marshy northern coast and narrow southern coast. It covers an area of 3.17 km^2, and its maximum (effective) length is approximately 4 km. In the past, Lake Ludas was a stagnant water body characterized by huge water-level fluctuations, which were the consequence of irregular periodic change of precipitation. The water regime of Ludas, as well as Palic, was entirely under the influence of meteorological factors, until the human factor became predominant.

There is a large amount of sediment deposited at the bottom of both lakes. The sediment thickness varies from 0.3 to 1.2 m, which additionally decreases the water column [27]. Consequently, the water depth of the (largest) sector 4 of Lake Palic is 1.9 m, while the mean depth of the whole lake is 2.3 m. The average depth of Lake Ludas is 0.9 m.

Urbanization for over 100 years led to changing their natural characteristics. Water regime regulation and introduction of complex drainage measures in wider area led to significant changes: natural wetlands were drained and ecosystems were changed. Nowadays, the main water source that recharges Lake Palic is wastewater treatment plant, which discharges around 13 million m^3 of water into the lake, whose quality satisfies the EU and Serbian standards [28]. The drawback of this system is that the volume of the lake is less than the discharged volume of treated wastewater during the year; therefore, there is no dilution in lake water, and nutrient load (mass) is huge. Lake Ludas also receives treated and untreated wastewater and runoff from agricultural plots that surround the lake. Cultivated areas with intensive farming surround the most part of the lakes' shores. There are also livestock farms without appropriate collection and treatment of wastewater in the surrounding area, as well as resident weekend facilities and rural settlements. All mentioned factors affect more or less the water quality in these lakes, but it is mostly influenced by water recharge [28–30]. Formerly, Lake Ludas received the biggest amount of water from the northern sandy terrain, by the Keres watercourse. Regulation of this watercourse changed water regime in the area. Today, Lake Ludas receives considerable amount of water from the Palic-Ludas canal, which connects these two lakes.

2.2. The Palic-Ludas canal

The Palic-Ludas canal was constructed with the purpose to carry treated wastewater from the wastewater treatment plant (WWTP), through parts of Palic and Ludas lakes to the Keres River. A part of this system, small bordering canal, was designed to carry untreated and treated wastewater from the WWTP to the Keres River, to avoid inflow of nutrient-rich water in lakes, but it has not been completed. The Palic-Ludas canal was sized to receive the flow of 3.0 m^3/s. The total length of the canal is 4.5 km. The width of the bottom of the canal is 1.5 m along the first 2.83 km and 2.5 m along the rest 1.67 km of the canal. The designed side slope is 1:2 and the bed slope is 0.3%. As stated by the regulation on the protection of the Special Nature Reserve "Lake Ludas", the Palic-Ludas canal is a part of the special nature reserve, designated as the area of the third degree of protection.

The Palic-Ludas canal carries a mixture of atmospheric water that is collected by open-canal network in the surroundings, treated and untreated wastewater and Palic's water. It would be expected that water that inflows Lake Ludas would be of better quality, due to self-purification abilities of water bodies. However, it is not the case, because the Palic-Ludas canal became an unofficial recipient for wastewater, which additionally increases mass nutrients and degree of eutrophication of Lake Ludas. A huge amount of unevenly distributed sediment at the bottom and wastewater discharge in the northern part of the lake are the most dominant eutrophication triggers [31].

3. Methods for water quality assessment

For the purpose of the study, water samples were collected during the period from November 2013 to March 2015, in total of 11 samples per location, on six locations (**Figure 2**). Water samples were collected at 20–50 cm beneath the water surface. Samples were transported in cool containers under 8°C and tested within 12–24 h. The analyses of collected samples have been done in laboratories of the Department of Ecological Microbiology, Faculty of Agriculture, University of Belgrade and Institute for the Development of Water Resources "Jaroslav Cerni". The analyses of easy variable parameters and microbial activity of samples were performed immediately after their receiving in laboratory.

Chemical analyses that include total suspended solids (TSS), total nitrogen (TN), phosphorous (TP) and total organic carbon (TOC) were determined according to the standard methods [32, 33].

Fecal contamination indicators (total coliforms, fecal coliforms and *Enterococcus*) were detected by the most probable number (MPN) method, using three tubes in each dilution. Presumptive test for total and fecal coliforms was done using the MacConkey (lactose) broth with inverted Durham tubes at 37°C for 48 h. Presumptive test of *Enterococcus* was done by using azide-dextrose broth (37°C/48 h) and the confirmation by inoculation on the bile esculin agar (37°C/24 h). Results are reported as MPN per 100 mL of water. The total number of aerobic heterotrophic bacteria was done using the meso-peptone agar (MPA), after incubation at

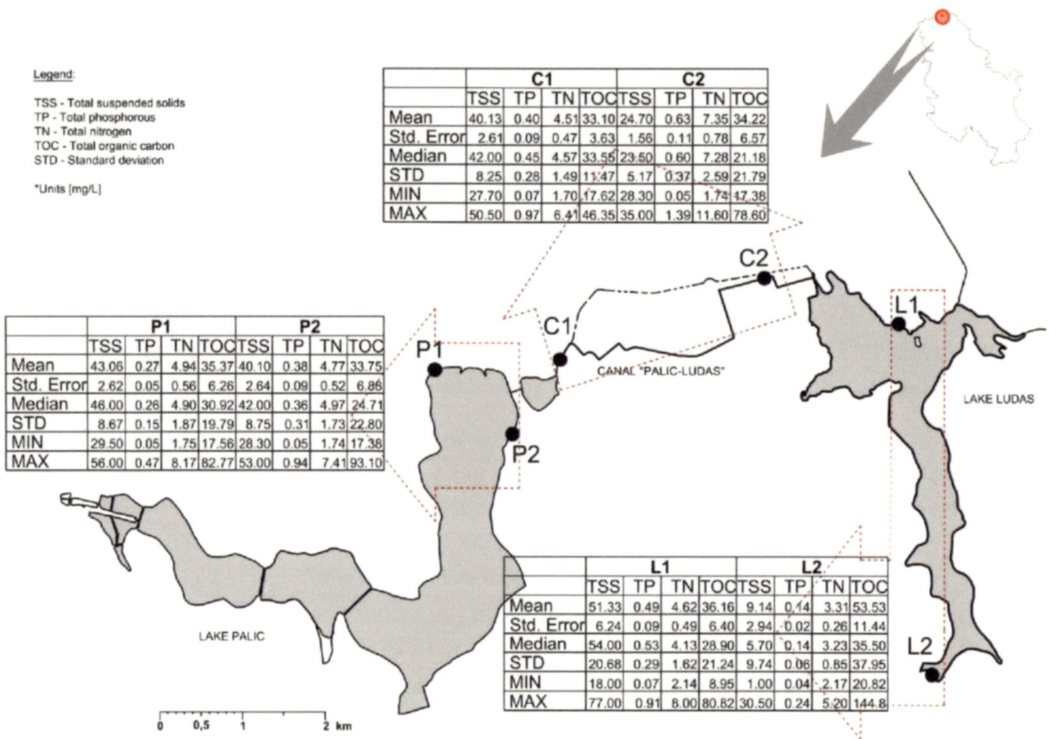

	C1				C2			
	TSS	TP	TN	TOC	TSS	TP	TN	TOC
Mean	40.13	0.40	4.51	33.10	24.70	0.63	7.35	34.22
Std. Error	2.61	0.09	0.47	3.63	1.56	0.11	0.78	6.57
Median	42.00	0.45	4.57	33.55	23.50	0.60	7.28	21.18
STD	8.25	0.28	1.49	11.47	5.17	0.37	2.59	21.79
MIN	27.70	0.07	1.70	17.62	28.30	0.05	1.74	17.38
MAX	50.50	0.97	6.41	46.35	35.00	1.39	11.60	78.60

	P1				P2			
	TSS	TP	TN	TOC	TSS	TP	TN	TOC
Mean	43.06	0.27	4.94	35.37	40.10	0.38	4.77	33.75
Std. Error	2.62	0.05	0.56	6.26	2.64	0.09	0.52	6.88
Median	46.00	0.26	4.90	30.92	42.00	0.36	4.97	24.71
STD	8.67	0.15	1.87	19.79	8.75	0.31	1.73	22.80
MIN	29.50	0.05	1.75	17.56	28.30	0.05	1.74	17.38
MAX	56.00	0.47	8.17	82.77	53.00	0.94	7.41	93.10

	L1				L2			
	TSS	TP	TN	TOC	TSS	TP	TN	TOC
Mean	51.33	0.49	4.62	36.16	9.14	0.14	3.31	53.53
Std. Error	6.24	0.09	0.49	6.40	2.94	0.02	0.26	11.44
Median	54.00	0.53	4.13	28.90	5.70	0.14	3.23	35.50
STD	20.68	0.29	1.62	21.24	9.74	0.06	0.85	37.95
MIN	18.00	0.07	2.14	8.95	1.00	0.04	2.17	20.82
MAX	77.00	0.91	8.00	80.82	30.50	0.24	5.20	144.8

Figure 2. Palic and Ludas lakes and their surroundings.

37°C during 48 h (mesophilic bacteria) and at 22°C during 3–5 days (psychrophilic bacteria). The amount of facultative oligotrophic bacteria was determined using 10 times of diluted MPA, after incubation at 22°C during 3–5 days. The results are expressed in CFU·mL^{-1}.

The determination of chlorophyll *a* concentration was done according to ISO 10260:1992. This method includes collection of algae by filtration, extraction of algal pigments and spectrometric determination of the chlorophyll *a* concentration in the extract.

The results were processed using MS Excel, R and RStudio [34].

4. Water quality: problem and targets

We used accustomed parameters to describe water quality of investigated water bodies. All samples showed rather high TOC values that indicate high organic matter loading; high values of chlorophyll *a* confirm the abundance of algae; and high contents of total nitrogen and phosphorous as well as TSS confirm previous statements.

The results show that the water quality of all water bodies significantly varies during the year (**Figure 2**) and that the values exceed limits for good ecological status, proposed by the Serbian regulation [35]. If we examine first the consequence of the lake state expressed by chlorophyll

http://dx.doi.org/10.5772/intechopen.74046

a, we notice that the content is high at every of four sampling lake locations (**Figure 3**). High content of chlorophyll *a* at the north of Lake Ludas (L1) is several times higher than other sampling locations of the same lake (L2). This fact indicates high intensity of biomass production in Lake Ludas. Even in late autumn, the content of chlorophyll *a* was extraordinarily high due to the clear weather and relatively high temperatures. Low chlorophyll *a* content at L2 doesn't mean that there is no nutrient loading, but that the submerged macrophytes retained its domination (**Figure 4**). Luxuriant submerged vegetation in the southern part of Lake Ludas fits very shallow water and soft lake bottom. Samples taken from Lake Palic at P1 and P2 do not show such differences (**Figures 2** and **3**).

Organic matter loading, expressed by TOC values, indicates intensive organic output (**Figure 2**). TOC values exceed four to five times of the proposed values (6 and 7 mg/L) for good ecological status of shallow lakes and artificial watercourses, respectively. The same is with TP and TN concentrations. In shallow temperate lakes, TP concentrations show more pronounced summer peaks with increase in trophic class, which is often attributed to increased concentrations of recharge water or increased internal loading [36]. The median values are mostly doubled, with the exception of samples taken at the south of Lake Ludas (L2). TSS and TP concentrations are satisfactory with respect to the Serbian regulation. The difference between north and south of Lake Ludas is obvious on many levels.

Water quality at the beginning (C1) and the end (C2) of the canal also differs significantly, with respect to most of parameters (**Figure 2**). Water at the beginning of the canal has properties similar to the water of Lake Palic, since the discharge of the lake in the canal is very close to sampling location. The content of different parameters at C1 is not permanently lesser than values obtained at the end of the canal. Average content of suspended matter is generally higher at C1. Also, higher TOC values are probably the consequence of high content of algae

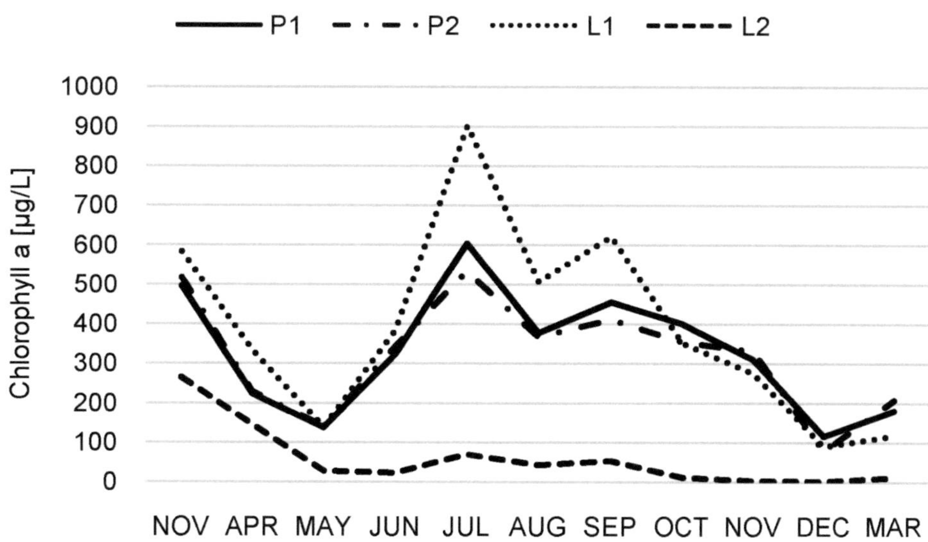

Figure 3. Lake states expressed by chlorophyll *a* during the year.

Figure 4. Lake Ludas (L1, left; L2, right).

in the water of Lake Palic that flows in the canal. Total phosphorus concentrations are in average of almost 10 times higher in the canal than the lake. The similar relation is with nitrogen concentrations (**Figure 2**).

These differences in chemical quality of water are confirmed by high fluctuation of bacterial counts during investigated period. After we analyzed separately each water body, we noticed some differences. Box plot diagrams (**Figure 5**) show the range of the obtained values, median and outliers. The range of registered values is wide, and the highest levels of fecal pollution indicators were registered in C2 samples.

We will use median for further commenting of obtained results rather than mean, since it gives a better idea of a "typical" value of each parameter. In Lake Palic, the registered number of bacteria was usually smaller at the site P2; particularly, median is notably lower for all indicator bacteria. At Lake Palic, total coliforms and enterococci exceed the limits of good ecological status in 18% of samples and fecal coliforms in 36% of samples. When considering just lake water, the worst sanitary quality is registered at L1 (Lake Ludas north). Fecal indicators are significantly higher in most samples taken from the L1. The range of observed values is wider, and median is higher for all examined parameters (**Figure 5**). Water at L2 was less bacteriologically polluted, regarding median values of indicator bacteria. This reveals that water from the Palic-Ludas canal has a considerable impact on the sanitary quality of water in the north of Lake Ludas, since the confluence of the canal is approximately 800 m away from the sampling point L1.

Majority of observed enterococci counts satisfied at least moderate ecological status (category III, in [35]) at all locations but C2. In 40% samples at C2, the ecological status with respect to this parameter was poor or unacceptable. The presence of coliform bacteria suggests that the canal receives fecal wastewater along the flow, which contributes to the reduction of effectiveness of self-purification process. Content of enterococci, as well as of total and fecal coliform bacteria, increases along the canal (**Figure 5**). The concentration range for enterococci at C2 is more expected for the raw sewage than surface water [37]. The origins of fecal matter in surface water could be organisms defecating in water, runoff from the surface and also discharge of poorly treated or untreated wastewater. Disposal of biological waste, as well as fertilization with manure, can in the same time increase the loading of nutrients (phosphorous, nitrogen) and coliforms in surface water [38].

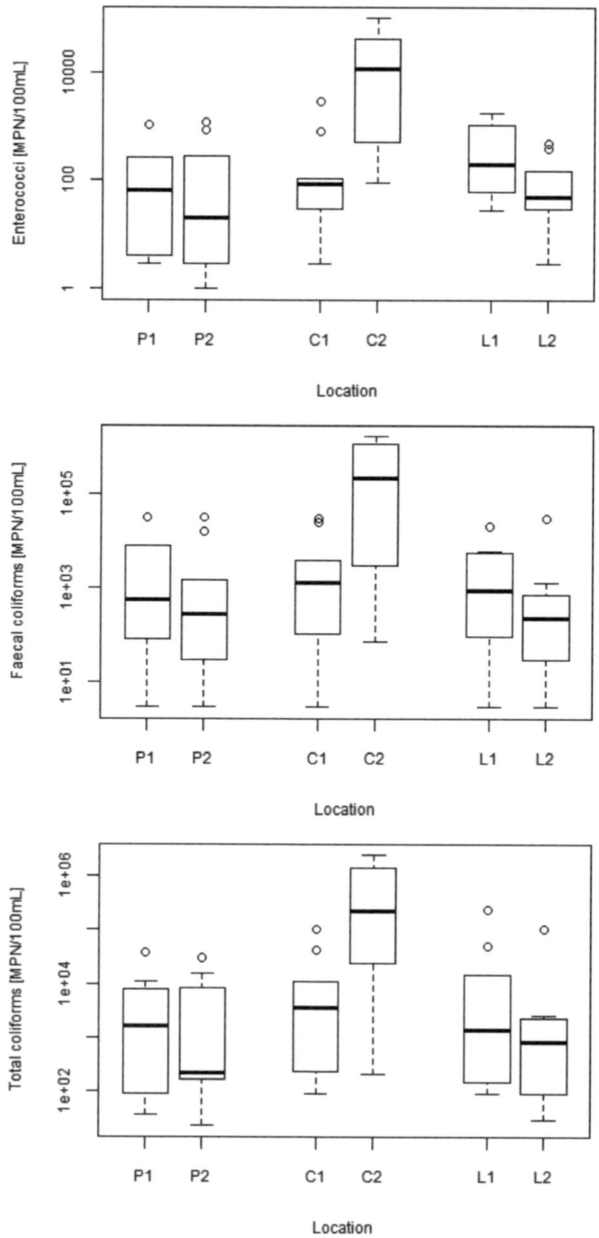

Figure 5. Microbial water quality.

For assessing self-purification ability, we used FO/H index [39]. Although this parameter is still not standard, Petrovic et al. [39] suggest using this parameter to obtain more in-depth ecological information about water quality. It represents the ratio between counts of facultative oligotrophic bacteria (FO) and heterotrophic bacteria (H). Accordingly, classification of water by self-purification ability is assessed as follows: FO/H < 1 pour, ≥1 satisfactory and >10

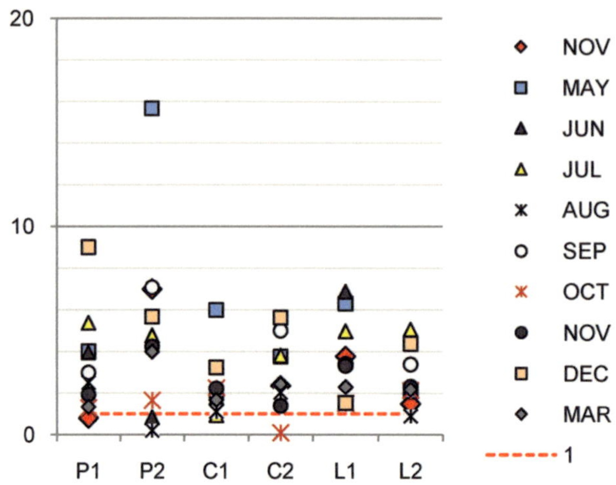

Figure 6. Self-purification ability expressed by FO/H index.

good. The self-purification process along the water flow, from Lake Palic through the canal to Lake Ludas, is satisfactory (**Figure 6**). However, the content of nutrients is still high, which indicates the existence of external sources.

Furthermore, finding widely distributed significant opportunistic pathogen of humans, animals and plants, *Pseudomonas aeruginosa* in Lake Palic and Palic-Ludas canal, but not in Lake Ludas itself [40], is in line with obtained self-purification capacity of water.

Differences in composition of saprophytic and potentially pathogenic microorganisms in different parts of the Lake Ludas can be explained not only by geographic features but also by quality of water that is transported by the canal into the northern part of the lake. Large organic load reflects also in a high (occasionally higher) number of saprophytic heterotrophic bacteria in comparison to indigenous oligotrophic bacteria, which reduces the self-purification ability and implies accelerated eutrophication of Lake Ludas and environmental threat.

Hence, Lake Ludas, a center of biological diversity, where numerously rare, endemic and relict species exist, deteriorates due to the processes of sedimentation and plant overgrowing. As a result of intensive sedimentation, reed migrates in other parts of the lake, while old reed beds deteriorate. Improvement of the state of this area is relevant for the local community as well as for improvement of the environment and tourism development [41]. Our target is to achieve good ecological status of water bodies (**Table 1**), using bioremediation technologies. The focus is on the canal, since the water quality decreases along its path.

Parameter	Shallow lake	Artificial water body
TN (mg/L)	2	2
TP (mg/L)	0.2	0.3

Table 1. Good ecological status (excerpt from [35]).

5. Alternatives for water quality improvement

Aquatic macrophytes have multiple effects on nutrient cycling in lakes. Interactions among macrophytes, sediment and water may result in increase or decrease in the nutrient concentrations of the water, depending on the growth phase. Changes in biogeochemical cycles and food chains lead to the loss of biodiversity and the survival of a species that have adapted to the changes. These tolerant species in the new, changed, environmental conditions become dominant and their activities continue to alter the conditions of the environment, leading to a reduction in the ecological integrity and biological diversity necessary for the normal functioning of the ecosystem [42].

The dominant community in wetland ecosystems is reed beds. Reed covered vast shoals of Lake Ludas, mainly not only in the coastal region but also in the middle of lake, where it built a number of islands. This mosaic distribution turf like reed is a fundamental feature of Ludas, of particular importance as a nesting place of many rare species of birds. These islands have a significant phytosanitary and phytofiltration role. However, once lush and thick reeds are now at the stage of extinction. Reed tends to spread to the south, in the narrower part of Lake Ludas, with less deposited sediment. The reed belt, once found only along the southern coast of the lake, now spreads and slowly overgrows the open water zone. In the past, floating flowering was present between the reeds, but today only species is *Lemna minor* is present [43].

Finding solution for the improvement of lake water quality can be distinguished by implementing different chemical or biological treatment. Maintenance of clear water state in shallow lakes includes a reduction of external loading, internal phosphorous loading and presence of stable submerged macrophyte communities [16]. Every method has advantages and disadvantages, and every lake has certain specificities, so there is no recipe how to "cure" a shallow lake. But it is certain that the quality of water that recharges the lake is the key factor for maintaining satisfactory water quality of the lake. There are numerous factors that influence the best technique, starting with climate, location, elevation, surroundings, origin of pollution, intensity of agricultural practice, slopes, etc. [14]. Usually, complex measures are necessary to deal with human-induced problems. However, it is necessary to consider a cheap, environmentally friendly and durable approach to solve water pollution problems.

Macrophytes have a strong reductive effect on resuspension; they can substantially reduce internal phosphorus loading in lakes [44]. Macrophytes can also promote N retention in lakes by enhancing denitrification and by taking up nitrogen from the sediment [45]. In addition, macrophytes can successfully perform remediation of deposited sediment. Even some sensitive agricultural crops used for sediment phytoremediation, such as mustard or lettuce, showed satisfactory germination rates (60–80%) [46].

Inflow of a large amount of nutrients that influence the bloom of cyanobacteria is almost impossible to control in current conditions. As we mentioned before, nutrient concentrations in water of the Palic-Ludas canal are higher than in Palic, which proves the additional water pollution in the canal and represents additional burden for Lake Ludas. While the study area is a part of protected zone and these lakes are only possible recipients for treated and untreated wastewater, the only way to mitigate impact of discharged water is by careful management of macrophytes.

5.1. How to use macrophytes?

Macrophytes are extensive consumers of nutrients and have a great role in suspended solid removal, which means that rhizospheric oxidation has a major effect in wastewater purification. Previous research [47] showed that aquatic plants have a photosynthetic capacity to transform molecules and make them available for rhizospheric microorganisms which contribute to self-purification. Also, this process introduces the basis of phytoremediation. The application of appropriate macrophytes could be the right choice for in situ remediation technology which is very acceptable for developing countries [48]. Just 1 hectare of planted indigenous macrophytes can decrease nitrogen content in water for up to 8 t/year and phosphorous for up to 1 t/year (**Table 2**) [27]. Often, there is no enough space to plant such a large area, so floating mats with macrophytes can be a practical solution for enhancement of water quality [49, 50]. The mechanisms of nitrogen and phosphorus removal depend on the macrophyte ability of adsorption, but nitrogen removal also has a very close connection with microbial processes like nitrification and denitrification [51].

Constructed wetlands might be effective in treating nutrient pollution as well as in restoration of lake ecosystems, keeping in mind that constructed wetland systems should be carefully designed and managed [28, 52, 53]. The use of macrophytes in the ecological restoration and water remediation contributes to the establishment of ecological balance, due to their effectiveness in the assimilation of nutrients and its role in creating the conditions for microbial degradation of organic matter.

Plant species	pH optimum	Distance (m)	Rooting depth (m)	Nitrogen content (%)	Phosphorous content (%)	Dry matter yield (t/ha)
Emerged plants						
Typha spp.	4–10	0.60	0.3–0.5	14	2	30
Scirpus spp.	4–9	0.30	0.6	18	2	20
Phragmites spp.	2–8	0.60	0.4	20	2	40
Juncus spp.	5–7.5	0.15	0.3	15	2	50
Carex spp.	5–7.5	0.15	0.2	1	0.5	5
Submerged plants						
Potamogeton spp.	6–10	0.3		2–5	0.1–1	3
Myriophyllum spp.	6–10	0.3		2–5	0.1–1	9
Ceratophyllum spp.	6–10	0.3		2–5	0.1–1	10
Floating plants						
Lemna spp.				6	2	20 t/year

Table 2. Biomass and assimilated quantities of nutrients by common plants [27].

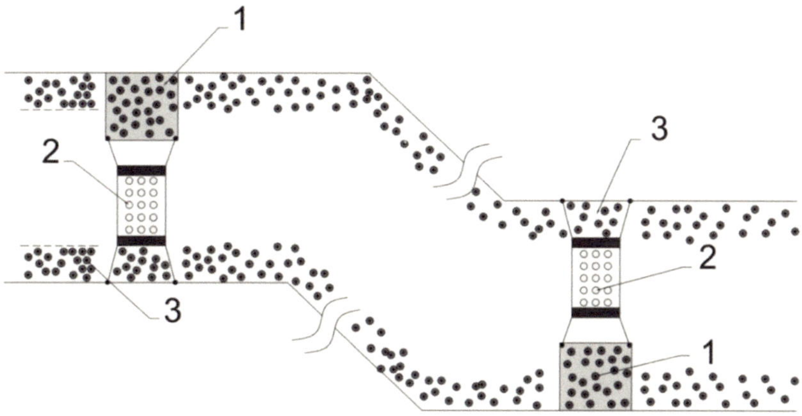

Figure 7. Bio-barriers placed alongside the canal: (1) reed on gravely substrate (biobank), (2) bio-bridge and (3) reed.

There are few solutions for improving the water self-purification ability in the canal. Usually, the good solution is to make more meanders, which increase the length of the water flow. It can be achieved by placing different types of barriers alternately alongside the canal (**Figure 7**). The study conducted by Lu et al. [54] showed that nutrient removal from aquatic ecosystem could become more effective by increasing hydraulic retention time. In this way microorganisms in water will have more time to decompose organic matter and release nutrients. Different plant species in canal will benefit from nutrients in readily available form. Joint activity between microorganisms and plants will contribute to the improvement of water quality. Open barriers in a form of small bridges along the canal will serve this purpose (**Figure 8**).

Bio-bridges consist of holders with a lot of small holes drilled in light plastic platform, floating at the water surface. In the platform holes, there are pots with chosen plants. In fact, they represent a form of floating wetlands that are fixed between canal banks. The plant roots will absorb different nutrients and bind suspended matter and microorganisms. As live system, root has a very large surface with a large amount of inhabited microorganisms that use different organic matters for their metabolic activities. This additional root biomass with a large number of different microorganisms will increase water self-purification rate.

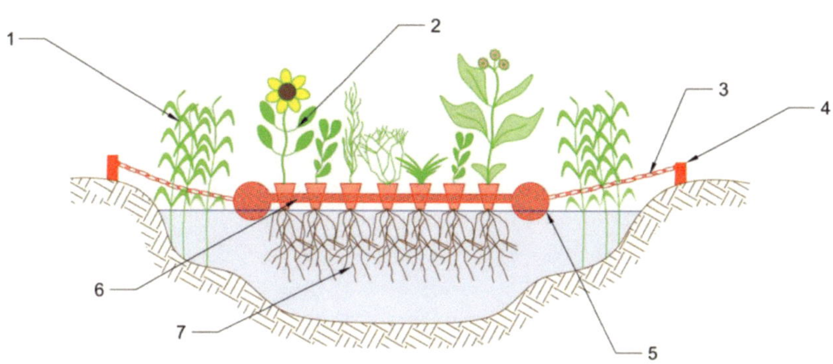

Figure 8. Bio-bridge: (1) reed, (2) different plants, (3) chain holder, (4) chain support, (5) floating carrier, (6) light platform with plants and (7) plant roots.

Generally, floating wetlands are more effective in comparison to other types. The large amount of nutrients is assimilated and eliminated from treated water, due to the removal of whole plant at the end of the growing season, not just above ground parts. Planted in this manner, macrophytes are forced to acquire their nutrition directly from the water column, which may enhance rates of nutrient and element uptake into biomass. This provides potential to enhance treatment performance by increasing the water depth retained during flow events to extend the detention time of storm waters in the wetland [50]. Plant species that can be used for floating wetlands, besides indigenous plants, are green salad, clover, alfalfa, mustards and sunflowers, although in that case, water below them should be aerated.

To improve self-purification, we suggest using a combination of bio-bridge and biobank, a kind of coarse sand/gravelly filter. This "filter" is in a steel net in a shape of parallelepiped that leans against the coast. Biobank performs coarse filtration of suspended matter. Moreover, it can be used as a substrate for planting autochthonous flora (**Figure 9**).

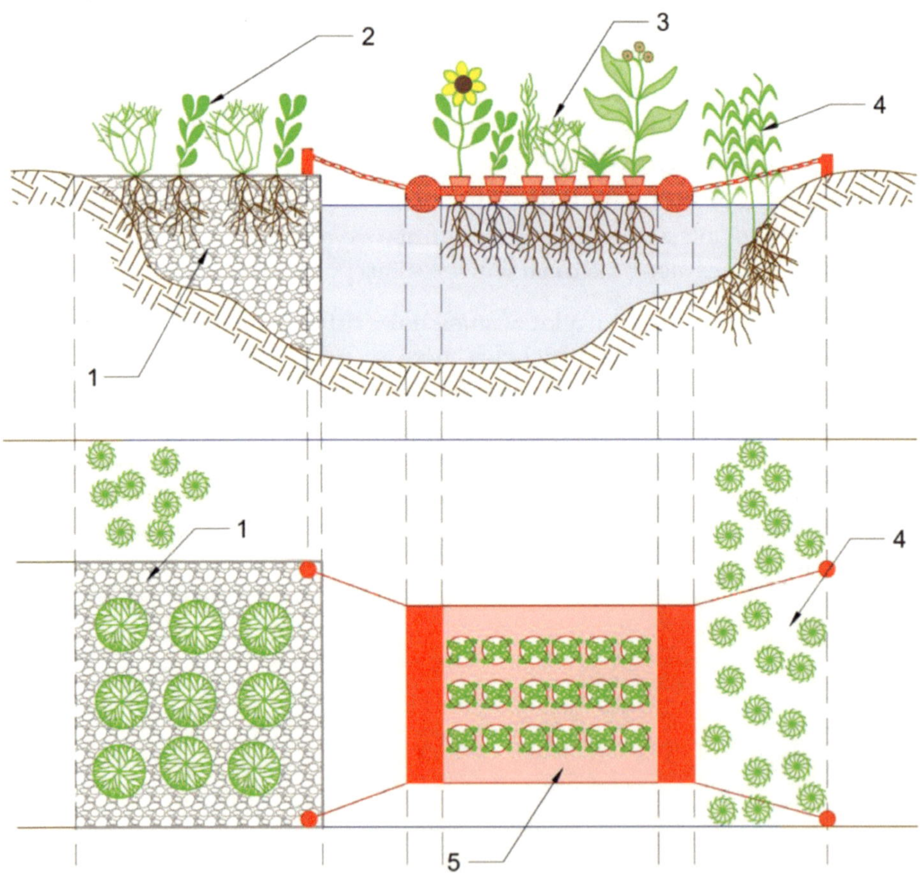

Figure 9. Biobank and bio-bridge: (1) gravelly substrate, (2, 3) different plants, (4) reed and (5) floating carrier of bio-bridge.

5.2. Sizing and application

The required area for future bio-bridges was calculated using average water quality at the end of the canal (**Figure 2**) and the required remedy level (**Table 1**) as input parameters. To achieve water quality that corresponds to good ecological status, it is necessary to remove about 64.470 kg TN/year and 3.980 kg TP/year. In addition, sizing was done according to efficiency of particular aquatic plants. Flora of lakes Palic and Ludas is characterized today by different aquatic marsh plants, such as *Phrarmites australis, Typha latifolia, Ceratophyllum demersum, Myriophyllum spicatum, Scirpus maritimus, Lemna minor*, etc. [43]. Plant composition for bio-bridges was selected accordingly (**Table 3**).

General drawback of constructed wetland application for restoration of surface water is their size. The canal has relatively small size, so the space for manipulation is limited. On the other hand, the flow should be slowed, not obstructed, so the obstacles have to leave enough space for water flow. Therefore, just 10 bio-bridges are anticipated for bioremediation. One hectare of planted macrophytes in this manner removes 6.075 kg TN/year and 730 kg TP/year or 9 and 18% of estimated total, respectively. Bio-bridges that cover much smaller area (120 m^2) assimilate only 9 kg TP/year and 73 kg TN/year, which is obviously insufficient for achieving planned targets. Biobank could stretch on maximum 0.8 ha, which would significantly improve efficiency of proposed system, but it might narrow down the cross-sectional area of the flow, i.e. the volume flow rate. Therefore, it is anticipated that biobanks cover an area of 0.4 ha. The places for installation of bio-bridges are proposed with respect to the change of water quality (deterioration along the water flow), nutrient content and allowed size (**Figure 10**).

Macrophytes assimilate considerable part of phosphorous from water, but some will precipitate, considering high water pH value. Together with the proven self-purification ability, it is possible to improve a certain extent of water quality that recharges Lake Ludas. Although it still seems to be insufficient for full treatment, it should be emphasized that bioavailable phosphorous which is incorporated in plant body will be permanently removed, and at least it will not make burden in the future by precipitating on the bottom sediment.

	Bio-bridges				**Biobanks**			
	Area	**DM***	**TN**	**TP**	**Area**	**DM***	**TN**	**TP**
	(m^2)	**(kg)**	**(kg)**	**(kg)**	**(m^2)**	**(kg)**	**(kg)**	**(kg)**
Typha spp. (30%)	36	108	15.1	2.2	1200	3600	504	72
Scirpus spp. (15%)	18	36	6.5	0.7	600	1200	216	24
Phragmites spp. (30%)	36	144	28.8	2.9	1200	4800	960	96
Juncus spp. (25%)	30	150	22.5	3.0	1000	5000	750	100
Total	120	438	72.9	8.8	4000	14,600	2430	292

*DM, dry matter.

Table 3. Nutrient removal by indigenous plants.

Figure 10. Locations for bio-bridges. They were set at 10 portions.

5.3. Advantages and disadvantages

The proposed system influences directly and indirectly the surroundings. Besides obvious advantage of proposed system of water treatment in environmentally friendly manner, it enhances biodiversity particularly when using autochthonous plants. Biobanks create heterogeneous habitats that are favorable for different macrophytes and for the growth and succession of other organisms, as well as for removing water pollutants. In sediment remediation study in [55], a general increase in microbial population was noticed in the lake sediment (2–5 times) after plant growth, which emphasizes the importance of phytoremediation. Inhabited bacteria play an important role in decomposition, and they are involved in various cyclic paths of different compounds in freshwater aquatic environments, which is particularly important for nutrient-rich aquatic environment. Semiaquatic coastal vegetation has also the anti-erosion role and affects sediment resuspension [56], which is of great importance since sediment resuspension has a strong effect on nutrient cycling of lakes [17].

The main disadvantage of this technology is occupation of large area, which is in this case a limiting factor. Consequently, bio-barriers for expected annual nutrient input seem to be undersized. In macrophyte management, one of the most important issues is the removal of plants (harvest). Plants absorb a significant amount of inorganic phosphorous, but after wilting nutrients are released in water again. Therefore, plant removal at the end of the growing season, before nutrient translocation, is significant for nutrient removal. In the absence of adequate machinery, the plant removal becomes labor-intensive.

Another advantage of bio-bridges is that they are completely removed at the end of the growing season, thus preventing the return of nutrients in the water. Produced biomass can further be used for different purposes like composting, energy production or building material.

6. Conclusions

Over decades the high concentration of nutrients, especially phosphorus, as well as change of hydrological balance in wider area, contributed to the disruption of ecological balance in the investigated shallow lakes, Lake Palic and Lake Ludas. Considering the input parameters of water quality and quantity, as well as target values, we proposed a model of constructed wetland, bio-barriers, within the artificial watercourse connecting two shallow lakes. It consists of several structures, so-called bio-bridges and biobanks, which maximize the potential of plants and microorganisms. Since limited action could be performed in protected areas, combination of different approaches for the improvement of their condition could be useful. In this way, macrophytes will contribute to restoration of biological structure and to the sustainable management of eutrophic water bodies. Despite spatial limitations, it is possible to remediate water to a certain extent, reduce the content of human pathogenic bacteria and remove nutrients with plant harvest.

Proposed concept is based on the plants that are tolerant to high levels of nutrients and predominantly indigenous species that grow in the surroundings of the canal and both lakes. This concept contributes, as the cost-effective and environmentally friendly technology, to the improvement of water quality, increase of biodiversity and maintenance of ecological balance in wider area.

Acknowledgements

This research was partially supported by the Ministry of Education and Science of the Republic Serbia (grant number TR31080).

Author details

Zeljka Rudic[1]*, Bojana Vujovic[1], Ljubinko Jovanovic[3], Dragan Kiković[4], Igor Kljujev[2], Mile Bozic[1] and Vera Raicevic[2]

*Address all correspondence to: zrudic@gmail.com

1 Institute for the Development of Water Resources "Jaroslav Cerni", Belgrade, Serbia

2 Faculty of Agriculture, Department for Microbial Ecology, University of Belgrade, Belgrade, Serbia

3 Educons University, Sremska Kamenica, Serbia

4 Faculty of Natural Science, Department for Biology, University of Kosovska Mitrovica, Kosovska Mitrovica, Serbia

References

[1] Dokulil M, Teubner K. Eutrophication and restoration of shallow lakes – The concept of stable equilibria revisited. Hydrobiologia. 2003;**506-509**:29-35

[2] UNEP. Global Environment Outlook 4 (GEO-4): Environment for Development. Nairobi: United Nations Environment Programme; 2007. Available from: http://pardee.du.edu/sites/default/files/GEO-4_Report_Full_en.pdf [Accessed: Nov 20, 2017]

[3] EEA. European Waters – Assessment of Status and Pressures. Copenhagen: European Environment Agency; 2012. Available from: www.eea.europa.eu/publications/european-waters-assessment-2012/at_download/file [Accessed: Nov 25, 2017]

[4] Jeppesen E, Søndergaard M, Zhengwen L. Lake restoration and management in a climate change perspective: An introduction. Water. 2017;**9**(122):1-8. DOI: 10.3390/w9020122

[5] Kearney MA, Zhu WX. Growth of three wetland plant species under single and multi-pollutant wastewater conditions. Ecological Engineering. 2012;**47**:214-220. DOI: 10.1016/j.ecoleng.2012.06.014

[6] Poikane S, Portielje R, van den Berg M, Phllips G, Brucet S, Carvalho L, Mischke U, Ott U, Soszka H, Van Wichelen J. Defining ecologically relevant water quality targets for lakes in Europe. Journal of Applied Ecology. 2014;**51**:592-602. DOI: 10.1111/1365-2664.12228

[7] Howarth RW, Ramakrishna K, Choi E, Elmgren R, Martinelli L, Mendoza A, Moomaw W, Palm C, Boy R, Scholes M, Zhu ZL. Nutrient management, responses assessment. In: Sarukhan J, Whyte A, editors. Ecosystems and Human Well-being 3, Policy Responses, Millennium Ecosystem Assessment. Washington: Island Press; 2005. pp. 295-311

[8] Pretty JN, Mason CF, Nedwell DB, Hine RE, Leaf S, Dils R. Environmental costs of freshwater eutrophication in England and Wales. Environmental Science and Technology. 2003;**37**:201-208. DOI: 10.1021/es020793k

[9] Jeppesen E, Søndergaard M, Meerhoff M, Lauridsen TL, Jensen JP. Shallow lake restoration by nutrient loading reduction – Some recent findings and challenges ahead. Hydrobiologia. 2007;**584**:239-252. DOI: 10.1007/s10750-007-0596-7

[10] Dodds WK, Bouska WW, Eitzmann JL, Pilger TJ, Pitts KL, Riley AJ, Schloesser JT, Thornbrughd DJ. Eutrophication of U.S. freshwaters: Analysis of potential economic damages. Environmental Science and Technology. 2009;**43**:12-19. DOI: 10.1021/es801217q

[11] Louette G, Declerck S, Vandekerkhove J, De Meester L. Evaluation of restoration measures in a shallow lake through a comparison of present day zooplankton communities with historical samples. Restoration Ecology. 2009;**17**(5):629-640. DOI: 10.1111/j.1526-100X.2008.00409.x

[12] Madgwick FJ. Restoring nutrient-enriched shallow lakes: Integration of theory and practice in the Norfolk Broads. UK: Hydrobiologia; 1999;**408/409**:1-12. DOI: 10.1023/A:1017058701974

[13] Wang GX, Zhang LM, Chua H, Li XD, Xia MF, Pu P. A mosaic community of macrophytes for the ecological remediation of eutrophic shallow lakes. Ecological Engineering. 2009;**35**:582-590. DOI: 10.1016/j.ecoleng.2008.06.006

[14] Moss B, Madgwick J, Phillips GA. A Guide to the Restoration of Nutrient-Enriched Shallow Lakes. London: Hawes; 1997. 180 p

[15] Meijer ML, Hosper H. Effects of Biomanipulation in the Large and Shallow Lake Wolderwijd. The Netherlands: Hydrobiologia; 1997;**342/343**:335-349. DOI: 10.1023/A:1017088701621

[16] Søndergaard M, Jeppesen E, Lauridsen TL, Skov C, van Nes EH, Roijackers R, Lammens E, Portielje R. Lake restoration: Success, failures and long-term effects. Journal of Applied Ecology. 2007;**44**:1095-1105. DOI: 10.1111/j.1365-2664.2007.01363.x

[17] Zhu M, Zhu G, Nurminen L, Wu T, Deng J, Zhang Y, Qin B, Ventela AM. The influence of macrophytes on sediment resuspension and the effect of associated nutrients in a shallow and large lake (Lake Taihu, China). PLoS One. 2015;**10**(6):e0127915. DOI: 10.1371/journal.pone.0127915

[18] Nurminen L, Horppila J. Life form dependent impacts of macrophyte vegetation on the ratio of suspended nutrients. Water Research. 2009;**43**:3217-3226. DOI: 10.1016/j.watres.2009.04.041

[19] Saaltnik RM, Dekker SC, Griffioen J, Wassen MJ. Vegetation growth and sediment dynamics in a created freshwater wetland. Ecological Engineering. 2018;**111**:11-21. DOI: 10.1016/j.ecoleng.2017.11.020

[20] Aboukila AF, Deng Z. Two variable residence time-based models for removal of total phosphorus and ammonium in free-water surface wetlands. Ecological Engineering. 2018;**111**:51-59. DOI: 10.1016/j.ecoleng.2017.10.014

[21] Vymazal J. Constructed wetlands for wastewater treatment: Five decades of experience. Environmental Science and Technology. 2011;**45**:61-69. DOI: 10.1021/es101403q

[22] Stefanik KC, Mitsch WJ. Structural and functional vegetation development in created and restored wetland mitigation banks of different ages. Ecological Engineering. 2012;**39**:104-112. DOI: 10.1016/j.ecoleng.2011.11.016

[23] Hammer UT. Saline lake ecosystems of the world. In: Monographiae Biologicae. Vol. 59. Dordrecht: Springer; 1986. 616 p

[24] Szilagyi F, Somlyody L, Herodek S, Istvanovics V. The Kis-Balaton reservoir system as a means of controlling eutrophication of Lake Balaton, Hungary. In: Jorgensen SE, Loffler H, editors. Guidelines of Lake Management. Vol. 3. Otsu: International Lake Environment Committee Foundation and United Nations Environment Programme; 1990. pp. 127-151

[25] Tatrai I, Matyas K, Korponai J, Paulovits G, Pomogyi P. The role of the Kis-Balaton water protection system in the control of water quality of Lake Balaton. Ecological Engineering. 2000;**16**:73-78. DOI: 10.1016/S0925-8574(00)00091-4

[26] Petrovic G. On the chemistry of some salt lakes and ponds in Yugoslavia. In: Williams WD, editor. Salt Lakes. Developments in Hydrobiology. Vol. 5. Dordrecht: Springer; 1981. pp. 195-200

[27] Raicevic V, Bozic M, Rudic Z, Lalevic B, Kikovic D, Jovanovic LJ. Eutrophication: Status, trends and restoration strategies of Palic Lake. In: Elshorbagy W, Chowdhury RK, editors. Water Treatment. Rijeka: InTech; 2013. pp. 355-380. DOI: 10.5772/50558

[28] Bozic M, Nikolic G, Rudic Z, Raicevic V, Lalevic B. Constructed wetlands as an alternative restoration measure for shallow lakes. Water Science and Technology. 2013;**68**(7):1672-1678. DOI: 10.2166/wst.2013.412

[29] Raicevic V, Bozic M, Rudic Z, Lalevic B, Kikovic D. The evolution of the eutrophication of Palic Lake (Serbia). African Journal of Biotechnology. 2011;**10**(10):1736-1744. DOI: 10.5897/AJBx10.028

[30] Rudic Z, Raicevic V, Bozic M, Nikolic G. Analysis of changes in water quality of a eutrophic shallow lake in the pannonian plaine – A modeling approach. In: Proceedings of the 9th Conference on Sustainable Development of Energy, Water and Environment Systems; 20-27 September 2014; Venice – Istanbul. Zagreb: Faculty of Mechanical Engineering and Naval Architecture; 2014. pp. 1-8 (Digital proceedings)

[31] Rudic Z. Factors of water quality degradation in shallow Pannonian lakes and their contribution to ecological risk [thesis]. Belgrade: University of Belgrade, Faculty of Agriculture; 2015

[32] APHA. Standard Methods for the Examination of Water and Wastewater, 14th ed. Washington, DC: American Public Health Association; 1975

[33] APHA. Standard Methods for the Examination of Water and Wastewater, 21st ed. Washington, DC: American Public Health Association; 2005

[34] R Core Team. R: A language and environment for statistical computing. Vienna: R Foundation for Statistical Computing; 2014. Available from: http://www.R-project.org/

[35] Regulation on limits of pollutants in surface and ground water and sediment and deadlines for their achievement. Belgrade: Official Journal of Republic of Serbia. 2012;**50/2012**. [in Serbian]

[36] Xie P. Biological mechanisms driving the seasonal changes in the internal loading of phosphorus in shallow lakes. Science in China: Series D Earth Sciences. 2006;**49**(1):14-27. DOI: 10.1007/s11430-006-8102-z

[37] Cabral JPS. Water microbiology, bacterial pathogens and water. International Journal of Environmental Research and Public Health. 2010;**7**:3657-3703. DOI: 10.3390/ijerph7103657

[38] Hofmann N, Beaulieu MS. A geographic profile of manure production in Canada, 2001. Catalogue no. 21-601-MIE-No.077. Ottawa: Statistics Canada; 2001. 20 p

[39] Petrovic O, Gajin S, Matavulj M, Radnovic D, Svircev Z. Microbiological Examination of Surface Water Quality. Novi Sad: Institute for Biology; 1998 [In Serbian]

[40] Vujovic B, Teodorovic S, Rudic Z, Bozic M, Raicevic V. Phenotypic heterogeneity of *Pseudomonas aeruginosa* isolates in the protected nature park 'Palic' (Serbia). Water Science and Technology – Water Supply. 2016;**16**:1370-1377

[41] Topalovic S. Current quality of the environment and possibilities of the development of sustainable tourism in the special nature reserve "The Ludas Lake". Bulletin of the Serbian Geographical Society. 2010;**XC**(2):189-198

[42] National Research Council. Clean Coastal Waters: Understanding and Reducing the Effects of Nutrient Pollution. Washington: National Academy Press; 2000. 428 p. DOI: 10.17226/9812

[43] Institute for Nature Conservation of Vojvodina Province. Nature Park "Palic" – Proposal for putting under protection as protected area of 3rd category. Nature protection study.

Novi Sad: Institute for Nature Conservation of Vojvodina Province; 2011. 234 p. [in Serbian]

[44] Horppila J, Nurminen L. The effect of an emergent macrophyte (*Typha angustifolia*) on sediment resuspension in a shallow north temperate lake. Freshwater Biology. 2001;**46**:1447-1455. DOI: 10.1046/j.1365-2427.2001.00765.x

[45] Christensen PB, Sørensen J. Temporal variation of denitrification activity in plant-covered littoral sediment from Lake Hampen, Denmark. Applied and Environmental Microbiology. 1986;**51**(6):1174-1179

[46] Radic D, Gujanicic V, Petricevic J, Raicevic V, Lalevic B, Rudic Z, Bozic M. Macrophytes as remediation technology in improving Ludas lake sediment. Fresenius Environmental Bulletin. 2013;**22**(6):1787-1791

[47] Brix H. Do macrophytes play a role in constructed treatment wetlands? Water Science and Technology. 1997;**35**:11-17. DOI: 10.1016/S0273-1223(97)00047-4

[48] Lu RS, Lo SL, Hu JY. Analysis of reservoir water quality using fuzzy synthetic evaluation. Stochastic Environmental Research and Risk Assessment. 1999;**13**(5):327-336. DOI: 10.1007/s004770050054

[49] Van de Moortel AMK, Laing GD, Pauw ND, Tack FMG. The role of the litter compartment in a constructed floating wetland. Ecological Engineering. 2012;**39**:71-80. DOI: 10.1016/j.ecoleng.2011.11.003

[50] Tanner CC, Headle TR. Components of floating emergent macrophyte treatment wetlands influencing removal of storm water pollutants. Ecological Engineering. 2011;**37**(3):474-486. DOI: 10.1016/j.ecoleng.2010.12.012

[51] Bednarek A, Szklarek S, Zalewski M. Nitrogen pollution removal from areas of intensive farming—Comparison of various denitrification biotechnologies. Ecohydrology & Hydrobiology. 2014;**14**:132-141. DOI: 10.1016/j.ecohyd.2014.01.005

[52] Li L, Li Y, Biswas DK, Nian Y, Jiang G. Potential of constructed wetlands in treating the eutrophic water: Evidence from Taihu Lake of China. Bioresource Technology. 2008;**99**:1656-1663. DOI: 10.1016/j.biortech.2007.04.001

[53] Vymazal J. Constructed wetlands for wastewater treatment. Water. 2010;**2**:530-549. DOI: 10.3390/w2030530

[54] Lu B, Xu Z, Li J, Chai X. Removal of water nutrients by different aquatic plant species: An alternative way to remediate polluted rural rivers. Ecological Engineering. 2018;**110**: 18-26. DOI: 10.1016/j.ecoleng.2017.09.016

[55] Petricevic J, Gujanicic V, Radic D, Lalevic B, Bozic M, Rudic Z, Raicevic V. The possibility of using macrophytes in Lake Palic sediment remediation. Archives of Biological Sciences. 2012;**64**(4):1481-1486. DOI: 10.2298/ABS1204481P

[56] James WF, Barko JW. Macrophyte influences on the zonation of sediment accretion and composition in a north-temperate reservoir. Archiv fuer Hydrobiologie. 1990;**120**(2):129-142

Chapter 9

Impact of Biochar on the Bioremediation and Phytoremediation of Heavy Metal(loid)s in Soil

Wenjie Sun, Sha Zhang and Chunming Su

Additional information is available at the end of the chapter

http://dx.doi.org/10.5772/intechopen.70349

Abstract

Anthropogenic activities, such as mining/smelting, result in the release and accumulation of heavy metal(loid)s in soil, posing serious human health and ecological risks. Due to the persistence of metal(loid)s, not undergoing any chemical and biological degradation, they can only be either immobilized or removed by, bioremediation and phytoremediation. Biochar is increasingly being recognized as a promising, effective material that can be used to remediate various contaminations including excessive heavy metals in soil. This chapter provides an overview of the state of the art on biochar resources, production processes and result of pyrolysis, surface characteristics of biochar, interactions of biochar with soil, and associated biota (microbes and plant). Furthermore, the understanding of characteristics of biochar and the interactions of biochar with soil and biota is necessary to assess the impacts of biochar on bioremediation and phytoremediation of heavy metal contaminated soil.

Keywords: biochar, heavy metal(loid)s, soil contamination, bioremediation, phytoremediation

1. Introduction

Soil contamination with heavy metal(loid)s can be attributed to many different sources such as agricultural and mining activities, industrial and residential release, which pose serious risks to environmental safety and human health. Various remediation techniques, based on either mobilization or immobilization processes, have been developed to solve these challenges. Of those methods, bioremediation, especially biochar amended phytoremediation, is increasingly being recognized as a promising technology that can be used to remediate various contaminations in soil. Many studies have reported that biochar has been effectively used to immobilize the metal(loid)s in contaminated soils and influence the bioavailability and bioaccessibility of metal(loid)s. Bioremediation, especially phytoremediation of metal(loid)s, has been extensively studied. Biochar, the carbon-rich product, is perceived to play significant roles on the bioaccessibility and bioavailability, hence biotransformation and bioremediation of heavy metal(loid) contaminated soil. However, biochar is often prepared from a variety of feedstocks under different pyrolysis systems, and thus, the surface properties could be varying significantly. It is urgent to address some key areas of research in order to ensure a safe and sustainable use of biochar. In particular, understanding the characteristics of various biochar, illustrating the effects of biochar properties on heavy metal(loid) behavior and transport, evaluating the influence of biochar on soil structure and nutrients, microbial community, plant growth, and consequently the effectiveness of bioremediation and phytoremediation are some of the areas that require attention.

This chapter provides an up-to-date review on the state of the art of environmental impacts and applications of biochar amended bioremediation and phytoremediation of heavy metal(loid) contaminated soil. In addition, the discussion will focus on the remediation of highly toxic metal(loid)s such as arsenic (As), cadmium (Cd), chromium (Cr), mercury (Hg), and lead (Pb) and different land-use scenarios that can potentially cause human health and phytotoxicity problems.

2. Characteristics of biochar

2.1. Physiochemical properties

Biochar is generally obtained from the low-temperature (e.g., 400–700°C) pyrolysis of various biomass such as manure [1], agricultural residues [2], bamboos [3], and biosolids in the absence of oxygen and is often used as a soil amendment [10], which is therefore distinguished from fuel charcoal. **Table 1** summarizes the physiochemical properties of biochar that is derived from wood material, agricultural by products, poultry litter, or sludge under different pyrolysis temperatures.

As shown in **Table 1**, although the physiochemical properties of biochar varied significantly due to the production from a variety of feedstocks under different pyrolysis systems, biochar is generally alkaline and has high surface area with large porosity, variable charges, and various functional groups. These characteristics can further affect its pH, cation exchange capacity (CEC), and surface sorption capacity. The particle size of biochar depends on the initial particle size of the feedstock, but it is likely to be smaller than the feedstock due to shrinkage and attrition during pyrolysis. An increased pyrolysis temperature could result in smaller biochar particles due to the decreased tensile strength of the feedstock at higher pyrolysis temperature [9].

Feedstock	Pyrolysis temperature (°C)	pH	CEC (mmol kg^{-1})	C (%)	C/N ratio	Total P (mg kg^{-1})	Ash (%)	Volatiles (%)	Surface area (m^2 g^{-1})
Oak wood [4, 5, 10]	60	3.7	182	47.1	444	5	0.3	88.6	n.d.
	350	4.8	294	74.9	455	12	1.1	60.8	450
	600	6.4	75.7	87.5	489	29	1.3	27.5	642
Pine needles [6]	300	6.4	n.d.	84.2	22	n.d.	7.2	38.6	4.1
	500	8.1	n.d.	90.1	22	n.d.	11.8	15.8	13.1
	700	10.6	n.d.	93.7	26	n.d.	18.7	6.2	391
Corn stover [4, 5, 10]	60	6.7	269	42.6	83	526	8.8	85.2	n.d.
	350	9.4	419	60.4	51	1889	11.4	48.8	293
	600	9.4	252	70.6	66	2114	16.7	23.5	527
Poultry litter [7]	0	8.2	n.d.	7.9	n.d.	n.d.	16.9	74.3	1
	350	8.7	n.d.	24	n.d.	n.d.	30.7	42.3	4
	700	10.3	n.d.	36	n.d.	n.d.	46.2	18.3	51
Poultry litter [10]	60	7.5	363	24.6	13	16,685	36.4	60.5	n.d.
	350	9.7	121	29.3	15	21,256	51.2	47.2	47
	600	10.3	58.7	23.6	25	23,596	55.8	44.1	94
Poultry litter [8]	450	9.9	n.d.	38	19	11,600	n.d.	n.d.	n.d.
	550	13til	n.d.	33	39	1800	n.d.	n.d.	n.d.
	300	5.3	n.d.	25.6	7.7	493	52.8	33.8	n.d.
Sludge [9]	400	4.9	n.d.	20.2	8.4	740	63.3	25.7	n.d.
	500	7.3	n.d.	20.3	10	568	68.2	20.7	n.d.
	700	12	n.d.	20.4	17	528	72.5	15.8	n.d.

Note: n.d. means not detected.

Table 1. Physiochemical properties of biochar derived from different feedstocks under different pyrolysis temperatures.

2.1.1. Alkalinity

The alkalinity of biochar is greatly affected by the type of feedstocks and pyrolysis temperature. It seems that biochar, derived from agricultural residuals and poultry letter in particular, is alkaline even at low pyrolysis temperature (<300°C), whereas biochar that derived from wood material is acid under low pyrolysis temperature [6, 8, 10]. The content of the carbonates are considered as the main alkaline components in biochar [11]. During the pyrolysis process, functional groups such as carboxyl and hydroxyl molecules on the surface of biochar are formed and also greatly contribute to the alkalinity of biochar. In general, biochar alkalinity increases with the increase in pyrolysis temperature [7, 11, 12] as shown in **Table 1**.

2.1.2. Nutrient

Biochar contains various nutrients such as P, K, Ca, and Mg, which are concentrated from the pyrolysis feedstock. Additionally, dissolved organic matter is formed during the pyrolysis process. Thus, the addition of biochar could provide a source of bioavailable nutrients to plants and microorganisms. However, the type and amount of bioavailable nutrients in biochar significantly depend on the feedstock type and pyrolysis conditions. When biochar is produced from poultry manure, peanut husk, and pine tree at 400 and 500°C, respectively, the contents of C and N varied greatly in biochar. In addition, the exchangeable P, K, Ca, and Mg were much higher in biochar obtained at 500°C than that at 400°C [13]. The difference was mainly attributed to the higher pyrolysis temperature that increased the mineralization of feedstocks, but decreased the CEC. From this perspective, it is most important to obtain nutrient-rich biochar from nutrient-rich feedstock under proper pyrolysis conditions. Generally speaking, plant-derived biochar has a relatively lower nutrient content than manure-derived biochar [14].

2.1.3. Stability

When biochar is added into the soil, it will present as isolated particulates, which distinguish from other forms of stable organic matter, either entrapped in soil pores and adsorbed to mineral surfaces or occluded in aggregates. Biochar containing more aromatic black carbons onto the surface is more persistent in the soil than any other form of organic carbon [15] and thus could enhance carbon sequestration in soils [2]. A previous study showed that the mineralization rate of biochar is very low, and the half-lives of carbon in biochar range from 102 to 107 years [16]. Another study has reported that fine biochar particles have remained in soils in humid tropical climates, such as in Amazon, for thousands of years [17].

2.2. Interactions between biochar and metal(loid)s

In the environment, biogeochemical reactions play important roles in influencing the fate, transport, and transformation of metal(loid)s. Since the ionic metal(loid)s may exist as cationic and anionic forms, the behavior of these ions will be affected through the interactions with positive and negative charges on the biochar surface. When mixed with topsoil, biochar with negative charges can tightly adsorb the cationic metal(loid)s such as Cd^{2+} and Pb^{2+} [18, 19], while biochar with positive charges can retain anionic metal(loid)s such as arsenite ($H_2AsO_3^-$) and arsenate ($H_2AsO_4^{3-}$). For cationic metal(loid)s such as Pb^{2+}, the physical adsorption, surface (co)precipitation, and surface/inner complexation with functional groups are considered as the major mechanism for the immobilization of Pb by biochar. Therefore, the biochar-induced changes in soils such as the increase of soil pH can further lower the solubility of cationic metal(loid)s. Because the physiochemical properties of biochar are a function of feedstock type and pyrolysis conditions (e.g., temperature and rate of temperature increase), it is necessary to determine suitable feedstock for biochar production that have the potential to remediate different metal(loid) in particular soils. On the other hand, unlike cationic metal(loid)s, anionic forms of metal(loid)s such as As ($H_2AsO_3^-$ and $H_2AsO_4^{3-}$), SeO_4^{2-} and Cr (CrO_4^{2-} or $HCrO_4^-$) are often dominant species in alkaline soils and weakly adsorbed by the negatively charged soil particles [20, 21].

It is well known that the oxidation state of metal(loid)s can affect the mobility in soil. For examples, the lower oxidation state of As (As(III) and As(V)) shows much higher mobility in soil, while the higher oxidation state of Cr (Cr(VI) and Cr(III)) presents much higher mobility in soils. Additionally, the redox potential in soils can alter the oxidation states of metal(loid) s. For instance, it was reported that biochar transformed Cr(VI) to the less mobile Cr(III) by continuously donating electrons, which may be associated with oxygen-containing functional groups on the surface of biochar [22]. Moreover, microbial metabolism by using biochar-derived organic carbon can also reduce Cr(VI) [16]. This reduction process resulted in the lower solubility of Cr and thus enhanced the immobilization of Cr in soil.

Soil pH and organic matter content also significantly control the adsorption/desorption processes of metal(loid)s in soils, where cationic metal concentrations are higher in acid soil solution. For example, Cr exists predominantly in the cationic form Cr^{3+} or $Cr(OH)^{2+}$ in acid soils (e.g., pH 3.5–6.0). The addition of biochar to soil can affect the pH and dissolved organic matter (DOC), and as a consequence to influence the mobility of metal(loid)s. Some studies revealed that biochar amended soil may increase the mobility of metal(loid)s such as As, Cu, and Sb [23–25]. For instance, the mobility of As was enhanced by the increase of pH in biochar amended soil [23, 26]. The possible mechanism for the mobilization is the electrostatic repulsion between anionic As and Sb species and negatively charged biochar surfaces, resulting in the desorption of As and Sb. In the case of Cu, the mobility is strongly correlated with the level of DOC in biochar. High DOC contents in biochar (pyrolysis temperature <500°C) can facilitate the organo-Cu complexes [27], but biochar with low DOC contents (e.g., pyrolysis temperature >600°C) will immobilize Cu by surface adsorption [25].

3. Bioremediation and phytoremediation with biochar

3.1. Bioremediation

Biochar application facilitates bioremediation of organic compounds [28–30]. The main mechanism is the increase of petroleum hydrocarbon-degrading microbial populations in biochar-amended soils [28]. Heavy metal(loid)s cannot be degraded or completely eliminated from the environment, but can be changed from one form to another, from high concentration to low concentration. Heavy metal(loid)s can also be accumulated in the biota. Therefore, there are two commonly used strategies for bioremediation of heavy metal(loid)s: (1) the absorption and accumulation to heavy metal(loid)s in woody plants and bioenergy crops in contaminated farmlands, and the removal of heavy metal(loid)s by harvesting the metal(loid)s accumulated biomass; (2) the transformation of heavy metal(loid)s into lower toxic products (e.g., complex state), and the adsorption by microorganism to reduce their toxicity and migration.

The potential benefits of biochar application in phytostabilization can be summarized in **Figure 1**. The negative charged surface of biochar and the alkaline nature can adsorb and retain toxic metal ions by different mechanisms. Biochar, at the same time, also provides a more favorable soil environment for beneficial microorganism and the development of plant roots by slowly releasing nutrients and maintaining good soil structures.

Microorganisms have been known to develop the resistance to toxic metal(loid)s and even evolved metabolic pathways to transform the metal(loid)s using several mechanisms. There are many factors affecting the effectiveness of bioremediation of heavy metal(loid)s. It has been reported that the reduction of Hg by a thermophilic *Streptomyces* species was associated with a novel Fe^{2+}-dependent mechanism in the membrane cytochrome c oxidase [31]. As we discussed before, the reductive precipitation of Cr(VI) to Cr(III) in soil was enhanced in the presence of biochar [22], which was due to the increased microbial activities promoted by the released carbon and nutrients from biochar [16]. In addition, dissimilative reduction may also facilitate immobilization of metal(loid)s such as Cr(VI) and U(VI) [32]; however, no reported evidences have illustrated the roles of biochar. Compared to the directly enhanced bioremediation, the presence of biochar could also provide indirect mechanism for the bioremediation of metal(loid)s. Microbial-induced calcite precipitation can strongly adsorb and/or coprecipitate the heavy metal(loid)s on its surfaces. During precipitation of calcite, heavy metal ions with ion radius close to Ca^{2+}, such as Cd^{2+}, Pb^{2+}, and Cu^{2+}, may be incorporated into the calcite crystal by substitution reaction [33]. Moreover, modern genetic engineering promotes the development of engineering microbes to remediation metal(loid) contaminated soils. For instances, genetically engineered *Ralstonia eutropha* has been reported to sequester Cd in soils and further reduce the availability of Cd [34]. Biochar facilitated

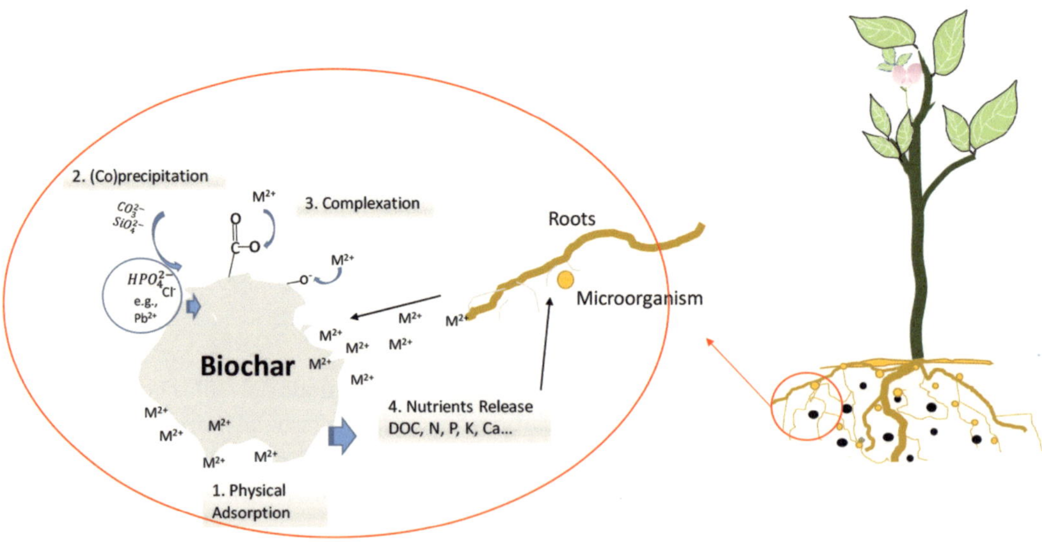

Figure 1. An overview of the potential positive effects attained by combining phytoremediation and biochar for bioremediation of heavy metal contaminated soil. Cationic metal ions expressed as M^{2+}. The potential benefits of biochar to phytoremediation are (1) physical adsorption of cationic metal(loid)s from soil pore water; (2) (co) precipitation with phosphate, carbonates, silicate, and chloride, e.g., the formation of pyromorphite; (3) complexation with functional groups on the surface of biochar; and (4) nutrients release such as N, P, K, Ca, and DOC. The processes (1)–(3) can reduce the bioavailable metal concentrations in soil pore water and further reduce the phytotoxicity. The process (4) can produce nutrients to plant roots and microorganism in rhizosphere.

this process by creating a microbe-favorable soil environment in the presence of biochar and thus potentially increased the bioremediation efficiency. Recently, a bamboo biochar oxygen-releasing bead (ORB) was reported to be a potential oxygen-releasing material for use in soil and groundwater bioremediation [35]. This type of biochar can potentially promote dissimilative oxidation of metal(loid)s such as As(III) to less mobile form.

3.2. Phytoremediation

Phytoremediation is a combined multidiscipline consisting of soil chemistry, microbiology, and plant physiology, aiming to (im)mobilize pollutants from various environmental media [36]. In general, phytoremediation includes phytoextraction, phytostabilization, phytodegradation, phytovolatilizationan, and rhizoremediation. In this chapter, phytoextraction and phytostabilization are discussed in detail. Compared with other remedial practices for heavy metal(loid)s, digging and dumping or chemical immobilization, phytoremediation is attracting more attention because of the effectiveness and low cost. Other benefits such as prevention of erosion or contaminant leaching are also very important for further land management and development. **Table 2** summarizes some biocar amended phytoremediation for heavy metal(loid)s contaminated soils.

Feedstock	Dose	Metal(loid)s	Effects	Plant species
Wood [51]	1, 2.5, and 5 %*w*/*w*	Ni, Mn, and Cr	Decrease exchangeable Ni, Mn, and Cr; improved plant growth	*Lycopersicon esculentum* L.
Hardwoods [26]	20% v/v	As	Increased pore water As but no effect on As uptake	*Miscanthus x giganteus*
Miscanthus [52]	1, 5, and 10% *w*/*w*	Cd, Pb, Zn	Decreased bioavailable metal concentrations and that in shoots but the biomass tripled	*Brassica napus* L.
Pruning residues; fir tree pellets and manure [43]	0, 1.5, and 3% *w*/*w*	Cd, Cr, Cu, Fe, Ni, Pb, Tl, and Zn	Decreased water extractable Cd, Cr, Cu, and Zn due to increased pH; No effect on Pb and Ni	*Anthyllis vulneraria, Noccaea rotundifolium and Poa alpina*
Oka, Ash, Sycamore, and Birch [53]	20% v/v	Cu and Pb	Decreased pore water Cu and Pb concentration and that in shoots	*Lolium perenne* L. var. Cadix
Hardwood [18, 24, 26]	–	As, Cu, Cd, and Zn	Increased pore water As and Cu; immobilization of Cd and Zn due to enhanced pH and DOC	*Solanum lycopersicum* L.
Sewage sludge [54]	0, 5, and 10% *w*/*w*	As, Cd, Cr, Co, Cu, Ni, Pb, and Zn	Decreased pore water As, Cr, Co, Ni, and Pb due to increase soil pH; mobilization of Cu, Zn, and Cd due to high available concentrations in biochar	Rice plants
Chicken manure and green waste [27]	1, 5, and 15% *w*/*w*	Cd, Cu, and Pb	Reduced NH_4NO_3 extractable Cd, Cu, and Pb; increased plant dry biomass by 353% for shoot; reduced Cd, Cu, and Pb accumulation by plants	Indian mustard (*Brassica juncea*)

Note: v/v means volume/volume and w/w means weight/weight.

Table 2. Biochar amended phytoremediation for heavy metal(loid) contaminated soils.

3.3. Phytoextraction

Phytoextraction, generally based on the hyper-accumulators or energy plants to uptake metal(loid)s, is the main method to remediate the metal(loid) contamination in soils. So far, hundreds of plant species have been used to accumulate As, Cu, Cd, Zn, Pb, Ni, Co, and Cr from soils. Ideally, plant species used for phytoextraction not only accumulate high amounts of the target metal(loid)s, but also have high biomass yield, tolerate the toxicity of the heavy metal(loid)s, be adaptive to soil and climate, be resistant to pests and diseases, and be easy to cultivate [37]. There are two factors as yield and the metal(loid) concentration determining the efficiency of phytoextraction. Hence, metal(loid) uptake, which is the product of these two parameters, can be either positive or null [38]. As far as we know, no studies meet all of the criteria mentioned above. However, there is a study reported that biochar-amended soil significantly improved the biomass of willow, yet the concentration of Cd and Zn in willow remained unchanged. Yet, phytoextraction is enhanced [39]. Phytoextraction in practice is often used in agricultural soils to lower the concentrations of toxic metal(loid)s below soil quality standards to improve soil environmental quality and to ensure food security. In highly contaminated soils, such as mine tailings, phytoextraction could cost hundreds of years. Thus, the contamination levels of the target agricultural soils should not be too high for phytoextraction. Although a hyper-accumulator could accumulate much higher metal(loid) concentrations, its slow growth rate often limits its application. Instead, energy plants such as sunflower and rapeseed plants are used to extract Cd from agricultural soils [40, 41]. Several studies have combined biochar and phytoextraction in practice. For instance, *Brassica napus L.* is used to extract Cd in the presence of biochar [42]. And different plant species and biochar have been also used in multicontaminated soils [43]. However, very few studies are available with a focus on the combination of biochar on phytoextraction of soil heavy metal(loid)s. For example, biochar amended phytoextraction was performed using *Amaranthus tricolor* L. to remediate the Cd contaminated soil [44]. Until now, most research results indicate that the addition of biochar reduces the bioavailability of heavy metal(loid)s, but plants require high concentrations of soluble metal(loid)s to extract and accumulate them. The benefits of biochar are to improve the characteristics of the polluted soil (water-holding capacity and nutrients) [45], enhance the soil microbial activity [10, 44–47], and have the potential to increase agricultural yields. Thus, using biochar to treat these soils not only immobilizes the heavy metal(loid)s but also increases the microbial biomass [48], which further lowers the ecological risk of heavy metal(loid)s in soil.

3.4. Phytostabilization

Phytostabilization is another phytoremediation process widely used for stabilization and containment of mine tailings [49]. The vegetative cover diminishes eolian dispersion and water erosion, while roots prevent leaching, which contribute to the immobilization of heavy metal(loid) s. The mechanisms involved in phytostabilization include precipitation, root sorption, complexation, or metal valence reduction. Phytostabilization, contrary to phytoextraction, primarily focuses on the heavy metal(loid) sequestration within the rhizosphere, but not in plant tissues. Metal(loid) stabilization is usually achieved through in situ application of soil amendments and microorganisms that effectively promote the metal immobilization and plant growth [42].

Most studies for biochar enhanced phytostabilization of metal(loid)s in soils include As, Cd, Cu, Cr, Ni, Zn, and Sb [25]. It is well known that As behaves differently from other metal(loid)s as a function of pH, since the As mobility can be reduced in acid soils due to increased adsorption on iron oxides under acidic conditions. Hartley et al. [26] showed that biochar can be used in combination with *Miscanthus* for phytostabilization. However, the results showed that the addition of hardwood biochar did not increase the As transfer to *Miscanthus* plants in three soils, whereas the alkaline biochar could mobilize As in soils. Cu and Pb were relatively easy to stabilize in biochar amended soil, while Cd and Ni varied greatly on the type of biochar added to soils [25]. The mechanism of stabilization is most likely due to the increase of soil pH. Extensive studies reported that soil amendments, such as lime, have the potential to be combined with phytoremediators so as to reduce the bioavailability of heavy metal(loid)s. By comparison, biochar seems to be more effective to regulating the availability of toxic elements and improving the biomass production of plants and remediation efficiency [39]. Improvements of plant yield with biochar addition are attributed to increased water and nutrient retention, improved biological properties and CEC, and buffed soil pH. Thus, biochar has a potential as an amendment for reducing metal uptake by plants. However, changes of soil pH in rhizosphere can potentially affect the effectiveness of biochar to immobilize metal(loid)s in soils and the acidification of the rhizosphere should be minimized [42].

It is perceived that biochar can interact with soil components for a prolonged period of time. As a result, the redox processes might result in the alteration of biochar, a process known as aging. The immobilization of heavy metal(loid)s by biochar was related to the metal lability (e.g., Pb^{2+} is prone to be immobilized first other than Cd^{2+}). During the aging process, a large variety of functional groups such as carboxylic, phenolic, and hydroxyl could be formed, and immobilization of cationic heavy metal(loid)s was not affected by biochar aging in soils with aged biochar [50].

4. Application of biochar to the bioremediation of mine area

Mining activities can cause the destruction of soil structure and change of soil biology and vegetation, resulting in extensive soil degradation. The contaminated soil after mining causes the devoid of vegetation due to metal toxicity and high acidity. Remediation and rehabilitation of these contaminated soils can be achieved via phytostabilization, a long-term and cost-effective rehabilitation strategy, promoting the revegetation to reduce the risk of pollutant transfer and ecological restoration [43], although these are difficult without proper soil amendments. The addition of biochar to contaminated soil may increase soil pH, water-holding capacity, and soil fertility, reduce the mobility of plant-available pollutant, and promote the revegetation [55]. Phytostabilization of mine tailings with biochar, produced from orchard prune residues and manure, at four different dosages (0, 1, 5, and 10%) showed the significant benefits of biochar to revegetate the plant species [43]. And the bioavailability of Cd, Pb, and Zn decreased proportionally with the increase of biochar content [43].

Biochar can reduce toxicity of heavy metal(loid)s and promote the plant growth in heavily contaminated soils [19, 56, 57], as shown in **Figure 1** (process 4). Soils with high available concentrations of toxic metal(loid)s are generally very acidic, unlikely supporting plant germination and growth.

4.1. Acid mine wastes

In highly multi-contaminants polluted soils, such as mine tailings, the phytoextraction process could take hundreds of years because of slow growth rate and low biomass of phytoextractors. Considering these limitations, revegetated phytostabilization, a long-term and cost-effective rehabilitation strategy, is one of the most promising remediation technologies to enhance metal stabilization in soils and reduce the risk of toxic metal(loid) transfer via leaching and water erosion [43]. However, when this technology is applied to mine tailings or contaminated soils, there are several issues that need to be addressed: (1) high metal(loid) contents cause high toxicity; (2) high acidity promotes metal release and enhances the toxicity; (3) the iron- and sulfur-oxidizing bacteria are dominated in acidic tailings that further increase acidity [58]; and (4) low nutrients cannot support plant growth. So far, no effective amendments have been developed to simultaneously fix these problems. Usually, the first step in practice is to add some liming materials such as limestone to increase alkalinity and increase the soil pH. Furthermore, the added alkalinity could be rapidly lost. And even if a higher soil pH is maintained, the lack of nutrients also makes the revegetation fail. However, the application of fertilizers to the lime-amended mine soils could further cause a higher acidity, which led to the rapid leaching out of added nutrients [59].

By comparison, biochar is one of the most relevant organic amendments. The addition of biochar to contaminated soil and waste rock piles can increase soil pH, water-holding capacity, and soil fertility, reduce the mobility of plant-available pollutants, and promote the revegetation [55]. Phytostabilization of mine tailings with biochar, produced from orchard prune residues and manure, at four different dosages (0, 1, 5, and 10%) exhibited the significant benefits to improve the revegetation of the plant species [43]. Additionally, the pH, nutrient retention, cation exchange capacity, and water-holding capacity of mine tailings increased, and the bioavailability of Cd, Pb, and Zn decreased proportionally with the increase of biochar content [43]. Soils developed on or around mine tailings are generally very acidic, with high available concentrations of dissolved toxic metal(loid)s, unlikely supporting plant germination and growth. Biochar is very stable with slow-releasing nutrients and readily reduces the plant-available concentrations of cationic metal(loid)s, resulting in the decrease of phytotoxicity of metal(loid)s to plants [19, 51, 56, 57], as the increase of the germination rate [60], root length, and better plant growth [51] are found in the presence of biochar.

4.2. Alkaline tailing soils

Alkaline tailing soils are another type of common metal(loid) contaminated soils, resulted from mining (e.g., Pb and Zn), smelting activities, or mining in arid regions [49]. These unclaimed or abandoned mining sites generally remain unvegetated for dozens or hundreds of years, and gradually, the contamination can greatly expand via eolian dispersion and water erosion [49]. The main concerns of high alkaline tailings and soils are that tailings and mining/smelting-affected soils become a significant source of air pollution, due to the formation of tiny particulate matter (e.g., PM 10 and PM 2.5), and as a consequence, they increase the exposure of heavy metal(loid)s to local residents. Therefore, it is urgent to improve the vegetation in alkaline tailing sites so as to control the release of particulate matter. Phytostabilization of mine tailings can be a promising strategy for metal immobilization in a long run, since vegetation will reduce the tiny metal-containing particles, while roots will prevent water erosion and immobilize metal(loid)s by adsorption or accumulation.

Although total metal(loid) contents in soils are very high (e.g., total Pb over 6000 mg kg^{-1} soil), the aqueous concentrations in soil solution (e.g., Pb below 10 μg l^{-1}) can be very low, due to the high soil alkalinity and the large fraction of nonchemical reactive metal(loid)s [61]. The addition of alkaline biochar will not improve the soil pH as it is already very high (e.g., 8), which reduce the solubility of metal(loid)s and prevent their bioavailability to plants.

However, because the replacement of topsoil with continuous addition of organic matter and nutrients is very expensive for extensive mine tailings sites, organic amendments such as biochar are commonly used as a substitute. Biochar can provide a slow-release fertilizer and serve as microbial carriers to improve soil quality, creating a favorable environment for further vegetation practices. In addition, biochar can improve soil structure, reduce erosion, and increase infiltration. Soil amendments such as biochar facilitate grassland recovery and revegetation in severely degraded habitats, such as mining waste and contaminated soils. For example, biochar reduced the toxicity of heavy metal(loid)s to plant growth in heavily contaminated soils [19, 56, 57]. Furthermore, biochar can improve soil structure and aeration and provide some necessary nutrients that promote microbial activity and diversity. Moreover, biochar can adsorb metal(loid)s on the surface and reduce the toxicity to microorganism. The improvement of soil enzymatic activities and microbial growth and diversities have been found in biochar-added soils [44, 46, 47, 51]. Biochar can alter soil microbial community, possibly including an increase in beneficial organisms that produce antibiotics, and can protect plants against pathogens [62]. However, some studies using wood-biochar as soil amendment do not show the same benefits [63].

5. Defects of biochar amended bioremediation

5.1. Phytoextraction of arsenic

From the perspectives of fate, transport, and transformation of the metal(loid)s, As behaves differently with other cationic metal(loid)s such as Cd. As exists predominantly as anionic form such as $H_2AsO_3^-$ and $H_2AsO_4^{3-}$ in soil solution. The mobility of As is pH-dependent and also greatly influenced by the biogeochemical interactions with soil minerals (e.g., iron oxides) in soils. The negatively charged As (As(III) and As(V)) are strongly adsorbed on iron oxides. However, when alkaline biochar is added to acidic or slightly acidic soils, the soil pH increased, which contributes to the increase of negative charges on the soil surface, promoting the release of adsorbed As into soil solution through charge repulsion. Several studies have reported that the As concentration in soil pore water increased with the increase of soil pH, induced by the presence of biochar derived from hardwood- or *Eucalyptus saligna* [18, 23, 24, 26, 64, 65]. Another intriguing factor that potentially increases the mobility of As is the introduction of dissolved organic matter from biochar. It is well known that the mobility of As and Fe also is greatly affected the by soil redox condition. An increase in dissolved organic matter can decrease the soil redox potential and enhance the reductive dissolution of solid iron oxides and release the adsorbed As to aqueous phase. In addition, the mobility of As(III) is higher than As(V). The reduction of As(V) to As(III) was predominantly (>83%) attributed to the biological reduction, stimulated by increased dissolved organic matter from biochar. The As(III) released from sediment upon biochar amendment (656 $\mu g\ l^{-1}$) was significantly higher than that from sediment without biochar addition (98 $\mu g\ l^{-1}$) after 49 days of cultivation [66]. Furthermore, the phosphate-extractable As in soil increased with biochar, because of the release of phosphate from biochar derived from hardwood [67]. Other anionic metal(loid)s, such as Sb, Cr, Mo, and Se, may also demonstrate the similar pattern as As in biochar-amended soils. To prevent the dissimilatory As reduction and mobility, a bamboo biochar oxygen-releasing bead (ORB) (104 days) was used to inhibit Fe(III) and As(V) reduction [35]. However, biochar derived from other feedstocks has not been well examined. Therefore, it is important to determine the impacts of biochar on the biogeochemical interactions that control the mobility of metal(loid)s in soil.

5.2. Phytoremediation of sewage sludge

Biochar seems to be very effective in controlling metal(loid) availability and facilitates the application of phytostabilization [68]. Sewage sludge, as an important feedstock for biochar production, is often used as soil amendments. However, there are concerns of pathogens and high available metal concentrations from sewage sludge and derived biochar. So far, the application of sludge-derived biochar is limited in both laboratory and field studies. The risk of using sludge in agricultural soils has still not been well evaluated, although there are reports that the concentrations of heavy metal(loid)s (i.e., Cu, Pb, Zn, Cd, and Cr) were concentrated in biochar, without increasing the metal uptake by Chinese cabbage [69].

Additional concerns raised for other metal(loid)-rich biochar, derived from plants used for phytoremediation or agricultural straw from contaminated soils. Although some studies claimed that metal(loid)s in feedstocks can mostly be fixed in biochar without the risk of leaching, the mechanism of fixation is still unclear. Pyrolysis temperature greatly affected the metal distribution in biochar product. With the increase of pyrolysis temperature, metal contents such as Zn and Pb decreased and perceived to form volatile metal chlorides [70]. However, the rest of metal(loid)s in biochar seem not increase the available metal(loid)s in soil in a 3-year field trails, even after repeated agricultural applications [71].

6. Future research

6.1. Characterization of biochar

Biochar has great potential as an amendment for phytoremediation, but its effect depends on the type of pyrolysis feedstock and temperature. Since most researches used different biochar produced from various feedstock, the impacts on the bioremediation varied greatly. However, these results are not comparable, and thus, a systematic evaluation is urgently needed for biochar-making procedure. The characteristics of the feedstock and pyrolysis conditions are crucial for the biochar selection. If the characteristics of feedstock can be qualitatively and quantitatively correlated with the properties of biochar, the selection and optimization of biochar becomes more promising [72]. The characterization of a range of surface properties of biochar and its biological effects on bioremediation are essential in order to implement these approaches effectively in practice [73]. The biochar-assisted phytoremediator targeting at different heavy metal(loid)s is strongly dependent on the interactions between specific metal(loid)s and biochar. How these properties are relevant for the adsorption of heavy metal(loid)s and how they contribute to the different mechanisms of heavy metal(loid)s immobilization need further examination.

6.2. Evaluation of biochar stability

Until now, the stability of biochar is still not well understood. The effectiveness of biochar on the improvement of soil quality has not been examined on a large timescale. In a 3-year field trail, the alkalinity associated with the biochar had been fully neutralized and most of the cations (i.e., K, Na, and Ca) from the biochar had been lost, but microbial community had built up [74]. The addition of biochar to soil causes small and potentially transient changes in a temperate agroecosystem functioning. However, most of these experiments for biochar amended bioremediation of heavy metal(loid)s (both under laboratory conditions and in field scale) have been conducted in the short term, which poses an interrogation on the long-term effectiveness. In fact, it could be expected that, due to aging processes, the ability and capacity of biochar to sequester heavy metal(loid)s decreases with time. More research is needed to understand the aging process in biochar. In addition, the degradation of biochar may not be avoided, and the degradation rate may vary greatly depending on feedstocks and pyrolysis conditions. It is still unclear which factors, e.g., feedstock type, pyrolysis conditions, or soil properties, determine the long-term stability of biochar. Thus, well-designed, large-scale and long-term field trials will be essential to evaluating the feasibility and stability of biochar amended bioremediation.

6.3. Elucidation of the mechanisms of biochar-assisted bioremediation

It is urgent to explore and improve the understanding of mechanisms involved in the biochar amended bioremediation of heavy metal(loid) contaminated soils. Biochar-induced changes in the soil ecosystem must be clarified before broad application, including the interactions among biochar, soil components, and soil microbial communities and plant roots. Until now, very few studies have been reported to systematically evaluate the joint application of biochar with phytoremediators under laboratory conditions or in field scale. For examples, although various approaches have been developed to remediate Pb contaminated soil, in which soil amendments with biochar for phytoremediation are the most suitable, practicable, and publically acceptable, but this practice still have some constraints due to the insufficient information to understand the immobilization processes. For other cationic metal(loid)s, biochar seem not work well in the phytoremediation of either acid or alkaline soils in some cases. In this case, without full understanding on the impacts of biochar on phytoremediators, this can result in an overestimation of heavy metal(loid) extraction and stabilization.

7. Conclusion

Biochar amended bioremediation is one of the important remedial technologies to remediate heavy metal(loid) contaminated soils. The biochar-enhanced phytoremediation has great potential to immobilize cationic heavy metal(loid)s in mine wastes and tailing soils, particularly those with high acidity. Biochar can reduce the bioavailability and leachability of cationic heavy metal(loid)s in soils, and improve the soil fertility and revegetation, and create a suitable environment for soil microbial diversity. However, the application of joint biochar and phytoextraction may be only capable of remediating the multicontaminated arable soils with low contamination levels, because biochar seems to be less effective to stabilize the potentially toxic cationic metal(loid)s such as Cd that have a high mobility in soils. In addition, it is important to select appropriate biochar so as to develop effective strategy to immobilize anionic metal(loid)s in situ. Furthermore, more thorough studies are needed to evaluate the efficiency of biochar amended bioremediation of highly contaminated alkaline soils. Future research should be performed with focuses on: illustrate the correlations among pyrolysis feedstocks, physic-chemical properties of biochar, and soil bioremediation; evaluate the biochar stability and its influence on fate and transport of metal(loid)s in mining tailings and soils on a large timescale; and understand the mechanisms of biochar-assisted bioremediation, especially involved in the interactions among biochar, soil particle, and soil microbial/plant roots, which is the key point for the development of cost-effective remediation strategies.

Author details

Wenjie Sun[1], Sha Zhang[1] and Chunming Su[2*]

*Address all correspondence to: su.chunming@epa.gov

1 Department of Civil and Environmental Engineering, Southern Methodist University, Dallas, TX, United States

2 National Risk Management Research Laboratory, Office of Research and Development, U.S. Environmental Protection Agency, Ada, OK, United States

References

[1] Cao XD, Ma LN, Gao B, Harris W. Dairy-manure derived biochar effectively sorbs lead and atrazine. Environmental Science & Technology. 2009;**43**:3285-3291. DOI: 10.1021/es803092k

[2] Woolf D, Amonette JE, Street-Perrott FA, Lehmann J, Joseph S. Sustainable biochar to mitigate global climate change. Nature Communications. 2010;**1**:56

[3] Xu T, Lou L, Luo L, Cao R, Duan D, Chen Y. Effect of bamboo biochar on pentachlorophenol leachability and bioavailability in agricultural soil. Science of the Total Environment. 2012;**414**:727-731

[4] Nguyen BT, Lehmann J, Hockaday WC, Joseph S, Masiello CA. Temperature sensitivity of black carbon decomposition and oxidation. Environmental Science & Technology. 2010;**44**:3324-3331. DOI: 10.1021/es903016y

[5] Nguyen BT, Lehmann J. Black carbon decomposition under varying water regimes. Organic Geochemistry. 2009;**40**:846-853. DOI: 10.1016/j.orggeochem.2009.05.004

[6] Ahmad M, Rajapaksha AU, Lim JE, Zhang M, Bolan N, Mohan D, Vithanage M, Lee SS, Ok YS. Biochar as a sorbent for contaminant management in soil and water: A review. Chemosphere. 2014;**99**:19-33. DOI: 10.1016/j.chemosphere.2013.10.071

[7] Cantrell KB, Hunt PG, Uchimiya M, Novak JM, Ro KS. Impact of pyrolysis temperature and manure source on physicochemical characteristics of biochar. Bioresource Technology. 2012;**107**:419-428. DOI: 10.1016/j.biortech.2011.11.084

[8] Chan K, Van Zwieten L, Meszaros I, Downie A, Joseph S. Using poultry litter biochars as soil amendments. Soil Research. 2008;**46**:437-444. DOI: 10.1071/SR08036

[9] Hossain MK, Strezov V, Chan KY, Ziolkowski A, Nelson PF. Influence of pyrolysis temperature on production and nutrient properties of wastewater sludge biochar. Journal of Environmental Management. 2011;**92**:223-228. DOI: 10.1016/j.jenvman.2010.09.008

[10] Lehmann J, Rillig MC, Thies J, Masiello CA, Hockaday WC, Crowley D. Biochar effects on soil biota: A review. Soil Biology & Biochemistry. 2011;**43**:1812-1836. DOI: 10.1016/j.soilbio.2011.04.022

[11] Yuan JH, Xu RK, Zhang H. The forms of alkalis in the biochar produced from crop residues at different temperatures. Bioresource Technology. 2011;**102**:3488-3497. DOI: 10.1016/j.biortech.2010.11.018

[12] Wu W, Yang M, Feng Q, McGrouther K, Wang H, Lu H, Chen Y. Chemical characterization of rice straw-derived biochar for soil amendment. Biomass and Bioenergy. 2012;**47**:268-276

[13] Gaskin J, Steiner C, Harris K, Das K, Bibens B. Effect of low-temperature pyrolysis conditions on biochar for agricultural use. Transactions of the ASABE. 2008;**51**:2061-2069. DOI: 10.13031/2013.25409

[14] Singh B, Singh BP, Cowie AL. Characterisation and evaluation of biochars for their application as a soil amendment. Soil Research. 2010;**48**:516-525

[15] Schmidt MW, Noack AG. Black carbon in soils and sediments: Analysis, distribution, implications, and current challenges. Global Biogeochemical Cycles. 2000;**14**:777-793

[16] Zimmerman AR. Abiotic and microbial oxidation of laboratory-produced black carbon (biochar). Environmental Science & Technology. 2010;**44**:1295-1301. DOI: 10.1021/es903140c

[17] Sombroek W, Ruivo MDL, Fearnside PM, Glaser B, Lehmann J. Amazonian dark earths as carbon stores and sinks [M]. In: Amazonian Dark Earths. Springer, Dordrecht; 2003. pp. 125-139

[18] Beesley L, Marmiroli M. The immobilisation and retention of soluble arsenic, cadmium and zinc by biochar. Environmental Pollution. 2011;**159**:474-480. DOI: 10.1016/j.envpol.2010.10.016

[19] Herath I, Kumarathilaka P, Navaratne A, Rajakaruna N, Vithanage M. Immobilization and phytotoxicity reduction of heavy metals in serpentine soil using biochar. Journal of Soils and Sediments. 2015;**15**:126-138. DOI: 10.1007/s11368-014-0967-4

[20] Zhang W, Zheng J, Zheng P, Tsang DC, Qiu R. Sludge-derived biochar for arsenic (III) immobilization: Effects of solution chemistry on sorption behavior. Journal of Environmental Quality. 2015;**44**:1119-1126

[21] McLaughlin MJ, Smolders E, Degryse F, Rietra R. Uptake of Metals from Soil into Vegetables. Springer, Dordrecht; 2011. DOI: 10.1007/978-90-481-9757-6_8

[22] Choppala G, Bolan N, Megharaj M, Chen Z, Naidu R. The influence of biochar and black carbon on reduction and bioavailability of chromate in soils. Journal of Environmental Quality. 2012;**41**:1175-1184. DOI: 10.2134/jeq2011.0145

[23] Beesley L, Moreno-Jimenez E, Gomez-Eyles JL. Effects of biochar and greenwaste compost amendments on mobility, bioavailability and toxicity of inorganic and organic contaminants in a multi-element polluted soil. Environmental Pollution. 2010;**158**:2282-2287. DOI: 10.1016/j.envpol.2010.02.003

[24] Beesley L, Marmiroli M, Pagano L, Pigoni V, Fellet G, Fresno T, Vamerali T, Bandiera M, Marmiroli N. Biochar addition to an arsenic contaminated soil increases arsenic concentrations in the pore water but reduces uptake to tomato plants (*Solanum lycopersicum* L.). Science of the Total Environment. 2013;**454**:598-603. DOI: 10.1016/j.scitotenv.2013.02.047

[25] Uchimiya M, Bannon DI, Wartelle LH, Lima IM, Klasson KT. Lead retention by broiler litter biochars in small arms range soil: Impact of pyrolysis temperature. Journal of Agricultural and Food Chemistry. 2012;**60**:5035-5044

[26] Hartley W, Dickinson NM, Riby P, Lepp NW. Arsenic mobility in brownfield soils amended with green waste compost or biochar and planted with *Miscanthus*. Environmental Pollution. 2009;**157**:2654-2662. DOI: 10.1016/j.envpol.2009.05.011

[27] Park JH, Choppala GK, Bolan NS, Chung JW, Chuasavathi T. Biochar reduces the bioavailability and phytotoxicity of heavy metals. Plant and Soil. 2011;**348**:439-451. DOI: 10.1007/s11104-011-0948-y

[28] Karppinen EM, Stewart KJ, Farrell RE, Siciliano SD. Petroleum hydrocarbon remediation in frozen soil using a meat and bonemeal biochar plus fertilizer. Chemosphere. 2017;**173**:330-339. DOI: 10.1016/j.chemosphere.2017.01.016

[29] Zhang HR, Tang JC, Wang L, Liu JC, Gurav RG, Sun KJ. A novel bioremediation strategy for petroleum hydrocarbon pollutants using salt tolerant Corynebacterium variabile HRJ4 and biochar. Journal of Environmental Sciences. 2016;**47**:7-13. DOI: 10.1016/j.jes.2015.12.023

[30] Garcia-Delgado C, Alfaro-Barta I, Eymar E. Combination of biochar amendment and mycoremediation for polycyclic aromatic hydrocarbons immobilization and biodegradation in creosote-contaminated soil. Journal of Hazardous Materials. 2015;**285**:259-266. DOI: 10.1016/j.jhazmat.2014.12.002

[31] Lloyd JR, Lovley DR. Microbial detoxification of metals and radionuclides. Current Opinion in Biotechnology. 2001;**12**:248-253

[32] Kashefi K, Lovley DR. Reduction of Fe (III), Mn (IV), and toxic metals at 100 C by *Pyrobaculum islandicum*. Applied and Environmental Microbiology. 2000;**66**:1050-1056

[33] Achal V, Pan X, Zhang D. Remediation of copper-contaminated soil by *Kocuria flava* CR1, based on microbially induced calcite precipitation. Ecological Engineering. 2011;**37**:1601-1605. DOI: 10.1016/j.ecoleng.2011.06.008

[34] Valls M, Atrian S, de Lorenzo V, Fernández LA. Engineering a mouse metallothionein on the cell surface of *Ralstonia eutropha* CH34 for immobilization of heavy metals in soil. Nature Biotechnology. 2000;**18**:661-665

[35] Wu CH, Chang SH, Lin CW. Improvement of oxygen release from calcium peroxide-polyvinyl alcohol beads by adding low-cost bamboo biochar and its application in bioremediation. CLEAN—Soil Air Water. 2015;**43**:287-295. DOI: 10.1002/clen.201400059

[36] Cunningham SD, Ow DW. Promises and prospects of phytoremediation. Plant Physiology. 1996;**110**:715

[37] Shakoor MB, Ali S, Farid M, Farooq MA, Tauqeer HM, Iftikhar U, Hannan F, Bharwana SA. Heavy metal pollution, a global problem and its remediation by chemically enhanced phytoremediation: A review. Journal of Biodiversity and Environmental Sciences. 2013;**3**:12-20

[38] Coumar MV, Parihar R, Dwivedi A, Saha J, Rajendiran S, Dotaniya M, Kundu S. Impact of pigeon pea biochar on cadmium mobility in soil and transfer rate to leafy vegetable spinach. Environmental Monitoring and Assessment. 2016;**188**:31. DOI: 10.1007/s10661-015-5028-y

[39] Brendova K, Tlustos P, Szakova J. Biochar immobilizes cadmium and zinc and improves phytoextraction potential of willow plants on extremely contaminated soil. Plant, Soil and Environment. 2015;**61**:303-308. DOI: 10.17221/181/2015-pse

[40] Jadia CD, Fulekar MH. Phytoremediation: The application of vermicompost to remove zinc, cadmium, copper, nickel and lead by sunflower plant. Environmental Engineering and Management Journal. 2008;**7**:547-558

[41] Ebbs SD, Kochian LV. Toxicity of zinc and copper to Brassica species: Implications for phytoremediation. Journal of Environmental Quality. 1997;**26**:776-781. DOI: 10.2134/jeq1997.00472425002600030026x

[42] Houben D, Sonnet P. Impact of biochar and root-induced changes on metal dynamics in the rhizosphere of *Agrostis capillaris* and *Lupinus albus*. Chemosphere. 2015;**139**:644-651. DOI: 10.1016/j.chemosphere.2014.12.036

[43] Fellet G, Marmiroli M, Marchiol L. Elements uptake by metal accumulator species grown on mine tailings amended with three types of biochar. Science of the Total Environment. 2014;**468**:598-608. DOI: 10.1016/j.scitotenv.2013.08.072

[44] Lu HP, Li Z, Fu SL, Mendez A, Gasco G, Paz-Ferreiro J. Combining phytoextraction and biochar addition improves soil biochemical properties in a soil contaminated with Cd. Chemosphere. 2015;**119**:209-216. DOI: 10.1016/j.chemosphere.2014.06.024

[45] Wang YJ, Li FY, Rong XM, Song HX, Chen JB. Remediation of petroleum-contaminated soil using bulrush straw powder, biochar and nutrients. Bulletin of Environmental Contamination and Toxicology. 2017;**98**:690-697. DOI: 10.1007/s00128-017-2064-z

[46] Ye J, Zhang R, Nielsen S, Joseph SD, Huang D, Thomas T. A combination of biochar–mineral complexes and compost improves soil bacterial processes, soil quality, and plant properties. Frontiers in Microbiology. 2016;**7**:372-384. DOI: 10.3389/fmicb.2016.00372

[47] Frankel ML, Bhuiyan TI, Veksha A, Demeter MA, Layzell DB, Helleur RJ, Hill JM, Turner RJ. Removal and biodegradation of naphthenic acids by biochar and attached environmental biofilms in the presence of co-contaminating metals. Bioresource Technology. 2016;**216**:352-361. DOI: 10.1016/j.biortech.2016.05.084

[48] Niemeyer JC, Lolata GB, de Carvalho GM, Da Silva EM, Sousa JP, Nogueira MA. Microbial indicators of soil health as tools for ecological risk assessment of a metal contaminated site in Brazil. Applied Soil Ecology. 2012;**59**:96-105. DOI: 10.1016/j.apsoil.2012.03.019

[49] Mendez MO, Maier RM. Phytostabilization of mine tailings in arid and semiarid environments: An emerging remediation technology. Environmental Health Perspectives. 2008;**116**:278

[50] Heitkötter J, Marschner B. Interactive effects of biochar ageing in soils related to feedstock, pyrolysis temperature, and historic charcoal production. Geoderma. 2015;**245**:56-64. DOI: 10.1016/j.geoderma.2015.01.012

[51] Bandara T, Herath I, Kumarathilaka P, Seneviratne M, Seneviratne G, Rajakaruna N, Vithanage M, Ok YS. Role of woody biochar and fungal-bacterial co-inoculation on enzyme activity and metal immobilization in serpentine soil. Journal of Soils and Sediments. 2017;**17**:665-673. DOI: 10.1007/s11368-015-1243-y

[52] Houben D, Evrard L, Sonnet P. Beneficial effects of biochar application to contaminated soils on the bioavailability of Cd, Pb and Zn and the biomass production of rapeseed (*Brassica napus* L.). Biomass & Bioenergy. 2013;**57**:196-204. DOI: 10.1016/j.biombioe.2013.07.019

[53] Karami N, Clemente R, Moreno-Jimenez E, Lepp NW, Beesley L. Efficiency of green waste compost and biochar soil amendments for reducing lead and copper mobility and uptake to ryegrass. Journal of Hazardous Materials. 2011;**191**:41-48. DOI: 10.1016/j.jhazmat.2011.04.025

[54] Khan S, Chao C, Waqas M, Arp HPH, Zhu YG. Sewage sludge biochar influence upon rice (*Oryza sativa* L.) yield, metal bioaccumulation and greenhouse gas emissions from acidic paddy soil. Environmental Science & Technology. 2013;**47**:8624-8632. DOI: 10.1021/es400554x

[55] Kelly CN, Peltz CD, Stanton M, Rutherford DW, Rostad CE. Biochar application to hardrock mine tailings: Soil quality, microbial activity, and toxic element sorption. Applied Geochemistry. 2014;**43**:35-48

[56] Younis U, Malik SA, Rizwan M, Qayyum MF, Ok YS, Shah MHR, Rehman RA, Ahmad N. Biochar enhances the cadmium tolerance in spinach (*Spinacia oleracea*) through modification of Cd uptake and physiological and biochemical attributes. Environmental Science and Pollution Research. 2016;**23**:21385-21394. DOI: 10.1007/s11356-016-7344-3

[57] Rehman MZU, Rizwan M, Ali S, Fatima N, Yousaf B, Naeem A, Sabir M, Ahmad HR, Ok YS. Contrasting effects of biochar, compost and farm manure on alleviation of nickel toxicity in maize (*Zea mays* L.) in relation to plant growth, photosynthesis and metal uptake. Ecotoxicology and Environmental Safety. 2016;**133**:218-225. DOI: 10.1016/j.ecoenv.2016.07.023

[58] Diaby N, Dold B, Pfeifer HR, Holliger C, Johnson DB, Hallberg KB. Microbial communities in a porphyry copper tailings impoundment and their impact on the geochemical dynamics of the mine waste. Environmental Microbiology. 2007;**9**:298-307. DOI: 10.1111/j.1462-2920.2006.01138.x

[59] Laird DA, Fleming P, Davis DD, Horton R, Wang B, Karlen DL. Impact of biochar amendments on the quality of a typical midwestern agricultural soil. Geoderma. 2010;**158**:443-449

[60] Ahmad M, Lee SS, Yang JE, Ro H-M, Lee YH, Ok YS. Effects of soil dilution and amendments (mussel shell, cow bone, and biochar) on Pb availability and phytotoxicity in military shooting range soil. Ecotoxicology and Environmental Safety. 2012;**79**:225-231. DOI: 10.1016/j.ecoenv.2012.01.003

[61] Zhang S, Song J, Gao H, Liu G. Assessment and modeling of Cd and Pb availability in contaminated arable soils in mining area of Guizhou. Soils. 2017;**49**:328-336. DOI: 10.13758/j.cnki.tr.2017.02.018

[62] Elad Y, Cytryn E, Harel YM, Lew B, Graber ER. The biochar effect: Plant resistance to biotic stresses. Phytopathologia Mediterranea. 2012;**50**:335-349. DOI: 10.14601/Phytopathol_Mediterr-9807

[63] Quilliam RS, Glanville HC, Wade SC, Jones DL. Life in the 'charosphere': Does biochar in agricultural soil provide a significant habitat for microorganisms? Soil Biology & Biochemistry. 2013;**65**:287-293

[64] Beesley L, Dickinson N. Carbon and trace element fluxes in the pore water of an urban soil following greenwaste compost, woody and biochar amendments, inoculated with the earthworm *Lumbricus terrestris*. Soil Biology & Biochemistry. 2011;**43**:188-196. DOI: 10.1016/j.soilbio.2010.09.035

[65] Ibrahim M, Khan S, Hao X, Li G. Biochar effects on metal bioaccumulation and arsenic speciation in alfalfa (*Medicago sativa* L.) grown in contaminated soil. International Journal of Environmental Science and Technology. 2016;**13**:2467-2474. DOI: 10.1007/s13762-016-1081-5

[66] Chen Z, Wang YP, Xia D, Jiang XL, Fu D, Shen L, Wang HT, Li QB. Enhanced bioreduction of iron and arsenic in sediment by biochar amendment influencing microbial community composition and dissolved organic matter content and composition. Journal of Hazardous Materials. 2016;**311**:20-29. DOI: 10.1016/j.jhazmat.2016.02.069

[67] Namgay T, Singh B, Singh BP. Influence of biochar application to soil on the availability of As, Cd, Cu, Pb, and Zn to maize (*Zea mays* L.). Soil Research. 2010;**48**:638-647

[68] Brendova K, Tlustos P, Szakova J. Can biochar from contaminated biomass be applied into soil for remediation purposes? Water, Air, & Soil Pollution. 2015;**226**:193-204. DOI: 10.1007/s11270-015-2456-9

[69] Liu T, Liu B, Zhang W. Nutrients and heavy metals in biochar produced by sewage sludge pyrolysis: Its application in soil amendment. Polish Journal of Environmental Studies. 2014;**23**:271-275

[70] Bert V, Allemon J, Sajet P, Dieu S, Papin A, Collet S, Gaucher R, Chalot M, Michiels B, Raventos C. Torrefaction and pyrolysis of metal-enriched poplars from phytotechnologies: Effect of temperature and biomass chlorine content on metal distribution in end-products and valorization options. Biomass & Bioenergy. 2017;**96**:1-11. DOI: 10.1016/j.biombioe.2016.11.003

[71] Lucchini P, Quilliam RS, DeLuca TH, Vamerali T, Jones DL. Does biochar application alter heavy metal dynamics in agricultural soil? Agriculture, Ecosystems & Environment. 2014;**184**:149-157

[72] Ashraf U, Kanu AS, Mo ZW, Hussain S, Anjum SA, Khan I, Abbas RN, Tang XR. Lead toxicity in rice: Effects, mechanisms, and mitigation strategies—A mini review. Environmental Science and Pollution Research. 2015;**22**:18318-18332. DOI: 10.1007/s11356-015-5463-x

[73] Wiszniewska A, Hanus-Fajerska E, Muszynska E, Ciarkowska K. Natural organic amendments for improved phytoremediation of polluted soils: A review of recent progress. Pedosphere. 2016;**26**:1-12. DOI: 10.1016/s1002-0160(15)60017-0

[74] Jones DL, Rousk J, Edwards-Jones G, DeLuca TH, Murphy DV. Biochar-mediated changes in soil quality and plant growth in a three year field trial. Soil Biology & Biochemistry. 2012;**45**:113-124. DOI: 10.1016/j.soilbio.2011.10.012

Chapter 10

Enhancement of Bioremediation and Phytoremediation Using Electrokinetics

Ikrema Hassan, Eltayeb Mohamedelhassan,
Ernest K. Yanful and Ze-Chun Yuan

Additional information is available at the end of the chapter

http://dx.doi.org/10.5772/intechopen.73202

Abstract

This chapter discusses the use of bioremediation and phytoremediation coupled with electrokinetics and presents the elements contributing to the success of the remediation process. A deep discussion and an overview of the current advancement in the biotechnologies are outlined in details. Innovative solutions for challenges facing the field application of the new technology are presented and new directions are proposed. A careful review for contaminated site conditions including pH, temperature, and other factors influencing the behavior of microbial community are presented. Great deal of discussion is around overcoming the adverse effect of electrolysis reactions, which is a by-product of electrokinetics. The discussion includes prolonging the survival of the indigenous bacteria, increase of microbial enzyme secretion, improvement of the indigenous bacteria metabolism, and exploration of metagenomics resources from soil biota. The challenges facing the field application of bioremediation and phytoremediation including the delivery of the electron donors and/or acceptors and nutrients to microorganisms involved in the biodegradation, particularly in clay soils, which has very low hydraulic conductivity, is discussed. The use of electrokinetics in biostimulation application to enhanced degradation of organic pollutant is reviewed. The implementation of bioaugmentation in bioremediation coupled with electrokinetics to enhance the outcome of bioremediation is presented.

Keywords: bioremediation, phytoremediation, electrokinetic, biostimulation, bioaugmentation

1. Introduction

In the last decade, different approaches have been used to mitigate soil contamination, however, many factors can affect the success of any remediation method including soil heterogeneity and the contaminant type. Researches and environmental engineers believe that there is no single remediation method that is suitable for every contaminated site; rather, an effective remediation program may involve the collective implementation of two or more methods [1]. Bioremediation is one of the most cost-effective remediation methods for contaminated soils [2]. Various bioremediation techniques have been used, with different degrees of success, to degrade pollutants at contaminated sites. Bioremediation types include biopile technique, landfarming approach, phytoremediation method, bioslurry, and bioventing.

Bioremediation employs microorganisms that have the ability to metabolize organic compounds including contaminants as food source in the soil. The traditional way to remediate contaminated sites typically depend on the type of soil and invariably involves "in situ" techniques like landfarming with occasional plowing or "ex situ" techniques such as windrows and biopile systems. In the former technique, the remediation process more or less relies on natural attenuation with minimal human input while in the latter cases, a great deal of human input and engineering is required to boost natural attenuation and accelerate remediation with minimal VOC emissions in the case of biopile systems. The literature covers a wide spectrum of approaches to soil remediation studies depending upon the discipline of the researchers: the treatment evaluation can be based on simple soil analysis for TPH, TOC (bulk parameters), or more sophisticated involving measurement of soil respiration rates and detailed chemical analysis of residual hydrocarbons in addition to the traditional bulk parameters. Indeed, recent studies indicate that relying on bulk parameters for the evaluation of the treatment process may still lead to highly hazardous residual petroleum hydrocarbon components [2, 3].

The success of a bioremediation process at specific site is mainly depend on the type of soil at the contaminated site, impermeable soil can exchange very little amount of oxygen and nutrients with the surrounding environment, therefore results on very slow remediation process. Microbiology area of research including microbial genomics, enzyme secretion, metabolism, and catalyst can be used to advance the knowledge in electrokinetic bioremediation. Bioremediation field applications are faced by obstacles such as the delivering of oxygen and nutrients to indigenous bacteria, particularly in soils with low hydraulic conductivity.

Electrokinetic remediation can be defined as the application of a low level direct current (DC) between a row of positively charged electrodes (anode) and negatively charged electrodes (cathode) placed at the edges of the soil under treatment [4]. The electric field incites three transport mechanisms, namely electroosmosis, electromigration, and electrophoresis, plus an electrolysis reaction at the electrodes. Electroosmotic flow is defined as the movement of water in the soil pores from anode to cathode under an applied electrical field. Electroosmotic flow (flow rate, q_A (m^3/s)) can be calculated using an empirical formula similar to Darcy's law of hydraulic flow rate:

$$q_A = k_e E\, A \tag{1}$$

where k_e (m^2/s V) is coefficient of electroosmosis permeability, E (V/m) is the electric field intensity $\left(-\frac{\Delta E}{\Delta L}\right)$, and cross-section area, A (m^2).

Electromigration is the transport of ions in the pores fluid toward the oppositely charged electrode. The migrational flux (J_m) (the ionic movement toward the oppositely charged electrode in soil pore fluid) can be calculated by Acar and Alshawabkeh [4].

$$J_m = c_j u_j \Delta(-E) \tag{2}$$

where effective ion mobility, u_j (m^2/s V), which is defined as the velocity of the ion in the soil under influence of a unit electric field gradient and can be evaluated as follows:

$$u_j = \frac{D_j z_j F}{RT} \tau n \tag{3}$$

where D_j (m^2/s) is the diffusion coefficient of ion species j in dilute solution, z is the valence of ion species j, F (96,487 C/mol) is the Faraday's constant, R (8.314 J/mol·K) is the universal gas constant, T (K) is the absolute temperature, τ is the tortuosity factor, and n is the porosity of the soil.

Electrophoresis is the movement of charged colloids under an applied electrical field. Electrolysis reactions produce hydrogen ions at the anode and hydroxyl ions at the cathode [5]. The hydrogen ions lower the soil pH near the anode and form an acid front, while the hydroxyl ions increase the pH at the cathode vicinity generating a base front. The acid front travels from the anode to the cathode, whereas the base front moves from the cathode to the anode.

Oxidation reaction at the anode:

$$2H_2O - 4e^- \rightarrow 4H^+ + O_2 \tag{4}$$

Reduction reaction at the cathode:

$$2H_2O + 4e^- \rightarrow 2OH^- + H_2 \tag{5}$$

Electrokinetic remediation is a timely technology that can significantly enhance nutrients delivery to indigenous bacteria, thereby providing a tremendous potential for cleaning contaminated soils including fine-grained soils, which are usually difficult to cleanup using conventional methods [6–8]. Many studies have investigated the use of electrokinetics to improve the outcome of bioremediation [3]. The combination of electrochemical technology with bioremediation may promote the removal of metal ions that are often inhibitory to bacterial activity, thereby enabling complete remediation of the soil [9]. Unlike pressure-driven flows in which channeling of the fluid through the largest pores is inevitable, electrokinetics permits a more uniform flow distribution and a high degree of control over the direction of the flow [10, 11]. Transport phenomena associated with electrokinetics can be utilized to effectively deliver nutrients to indigenous bacteria in the soils, and to enhance bioavailability (electroosmotic flow can enhance desorption).

Figure 1 shows the conceptual model where upon a release of petroleum hydrocarbon part of the contaminant evaporates and part moves through soil to contaminate subsurface. Depending on the site environmental conditions, indigenous microorganisms start to adapt and degrade

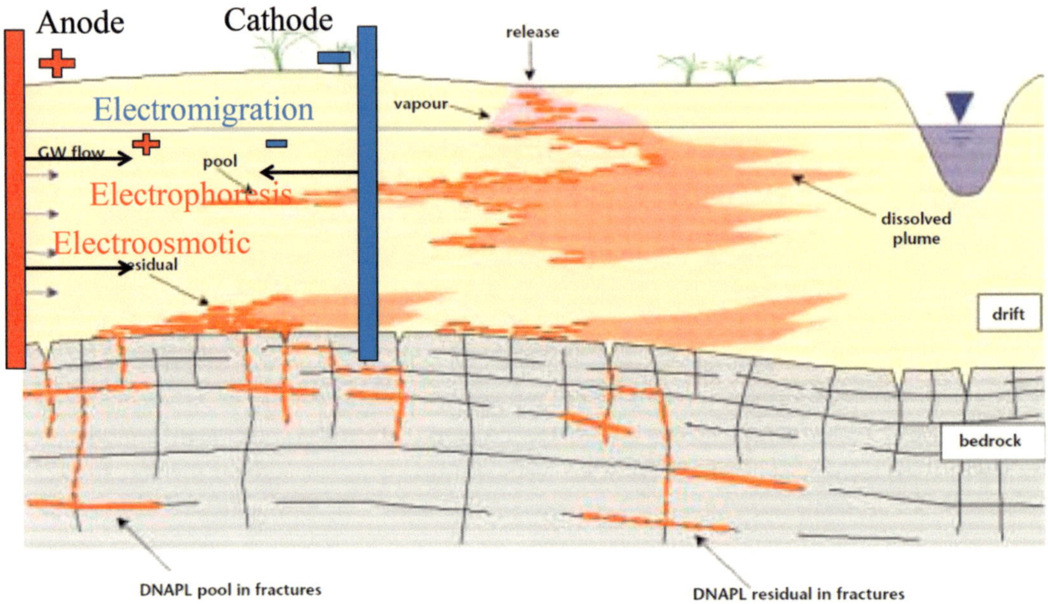

Figure 1. Conceptual model (after [43]).

contaminant. In most cases, the biodegradation rate is slow. Electrokinetics can be used to enhance the degradation rate by using the transport mechanisms associated with electrokinetics to deliver nutrients and/or to introduce new bacteria if the indigenous microorganisms are not capable of degrading the contaminant. However, the development of an acidic medium near the anode and an alkaline environment near the cathode by electrolysis reactions can create unfavorable condition for bacteria [7, 12, 13].

It is known that the application of the electric current during electrokinetic bioremediation increases the contaminated soil's temperature to high level which has an adverse effect on the survival of the indigenous bacteria. A recent study showed that the cost of electric power need for electrokinetic is a major part on the overall cost of electrokinetic remediation process. On the other hand, electricity power line may not be available in remotely located contaminated sites. Therefore, the energy consumption can increase the cost of the bioremediation process and result in restricting wide field applications of the electrokinetic bioremediation. Application of electrokinetic bioremediation can be divided into two main aspects:

- Microorganism-related factors, mainly how the environment at contaminated site can affect the degradation process.
- Electrokinetic processes; the phenomena associated with electrokinetic such as electrolysis reactions and the effect of other factors including the application of electric current, pH, soil temperature, and the availability alternatives concerning power source. In the following sections, each of the above-mentioned categories will be discussed.

http://dx.doi.org/10.5772/intechopen.73202

2. Microorganisms-related factors

The environment conditions at contaminated site directly affect electrokinetic bioremediation process through supporting the existence of microorganisms, or enhancing bacterial viability and persistence in contaminated soil. The subsurface conditions and the characteristics of indigenous bacteria are the main factors affecting electrokinetic bioremediation process. Several factors can contribute to persistence of bacteria in the soil environment including pH, nutrients, electron acceptors, osmotic stress, and temperature [14, 15]. In addition to soil environment, the characteristics of the indigenous bacteria play a significant role in their existence, for example, some bacteria form biofilms which protects themselves from external stresses [16, 17]. Another mechanism for bacteria to survive is to produce spores [18]. In the event of severe weather and nutrient deprivation, bacteria will die eventually producing endospores which have a very hard shell that protects them [19]. Typically, under extremely poor living conditions, endospores are in a state of dormancy (sleeping condition); once the environmental conditions improve, the spore will germinate and outgrow [20]. The following techniques can be used to enhance bioremediation:

1. Genetic engineering can be used to enhance the outcome of bioremediation through changing or rewiring of microorganism metabolic pathways so as to enhance the ability of bacteria in degrading the targeted contaminant [21, 22]. Also, modifying bacterial genes and regulatory networks to make them tougher to survive and tolerate high contaminant concentration in soils [23].

2. Enhancing bioremediation using a commercially available enzyme which is currently very expensive, but the expected reduction in enzyme prices as the technology continues to improve can significantly reduce the initial cost [24, 25]. Previous studies showed that the use of enzyme in bioremediation is not feasible because of the existence of heavy metals and other compounds which can denature the enzymes. It was concluded that the use of enzyme in bioremediation is not recommended because the enzyme will be denature under the field conditions and will not last long. An alternative way would be the use of microbial genetics to clone/overexpress and introduce exogenous gene/gene clusters using as a host bacterial cell that will produce the required enzyme. The other challenge will be the delivery of the enzyme expressed in bacteria to the contaminated zone in order to degrade the pollutant [26]. A recent study showed that the use of genetically modified bacterial secretion system can enhance the bioremediation [27].

3. Exploring metagenomics to enhance electrokinetic bioremediation [28]. It is known that only 5% of the microorganisms in nature are culturable in the laboratories and can be studied. Microorganisms in nature can produce enzymes that are capable of degrading contaminants. The total genomic DNA can be collected from contaminated soil, then using the available technologies to slice the DNA into pieces, and clone it into a vector to in order to form a library called 3D metagenomics DNA library. This library can be augmented into a selected bacterium that can be used to degrade contaminates providing that this library contains specific genes that can produce enzyme needed for the degradation process [29].

4. Characterizing the bacterial metabolism. It is very important to study the bacteria well before using the bacteria in the bioremediation process. In the contaminant degradation, knowing the degradation pathways are essential because some bacteria can produce secondary by-products/metabolites that are not in favor of the process. The knowledge of bacterial metabolism and the degradation pathways well can use to enhance the degradation by change/rewire bacterial metabolism to make them produce less/nontoxic products. More research is required in this area using the advancement in genetic engineering to eliminate/reduce the harmful by-products. Bacteria consortia at environmental sites communicate and coordinate behaviors and functionalities at community level using chemicals such as acyl homoserine lactones (AHLs), which is also described as bacterial quorum sensing (QS). In nature, QS may help bacteria for better and enhanced bioremediation. However, very few studies have investigated the role of QS in bioremediation [30–32].

3. Biostimulation and bioaugmentation

In bioremediation, there are three techniques that can be used at contaminated sites: (a) natural bioattenuation in which the pollutant is degraded by bacteria to nontoxic product, the process involve minimal to no human intervention; (b) biostimulation is defined as the introduction of supplements including nutrients, water, electron acceptors/donors to enhance the degradation process; and (c) bioaugmentation, which involves the use of large numbers of the bacterial strains that are capable of degrading the contaminant [33]. The following sections present the application of electrokinetics in biostimulation and bioaugmentation.

3.1. Biostimulation

In bioremediation, bacteria is employed to degrade contaminants into nontoxic products, the success of the degradation process is mainly dependent on the growth and reproduction of bacteria. The growth and metabolisms of microorganisms are affected by the available amount of nutrients, electron acceptors/donors, and oxygen. Electroosmotic can be used to deliver nutrients and oxygen to indigenous bacteria in the contaminated zone. Moreover, electrokinetics can increase the opportunities of contact between bacteria and contaminants therefore, enhancing bioavailability. Previous studies discussed the use of electrokinetics in delivering nutrients are presented in **Table 1**. In a previous study, electrokinetics was used to deliver nutrients to contaminated zone under controlled pH conditions [34]. In this study, polarity exchange technique was used to control the soil pH. The result showed that high amount of contaminant was degraded under controlled pH conditions. Nitrate and ammonium were delivered to contaminated zone under uncontrolled pH conditions. The results showed that nitrate can be delivered successfully/precisely to the contaminated area near the anode while low amount of ammonium was transported to the cathode. More importantly, it has been observed that the nutrients were distributed evenly in the soil under controlled pH using the exchange polarity technique [35]. The results from electrokinetic bioremediation studies have shown that electrokinetics is successful in delivering nutrients to indigenous bacteria.

Soil	Voltage/current	Nutrient concentration	Highlights/main outcome	Reference
Clay loam	1 V/cm	2 g/L NH_4NO_3 2 g/L KH_2PO_4	Nitrate transport rate 19 cm/d/v Phosphate results is not presented	[35]
Coarse sand	0.25 V/cm	1 g/L $NaNO_3$	Nitrate transported 0.6 cm/h	[41]
Clayey silt	0.5 V/cm	2 g/L NH_4NO_3 5 g/L KH_2PO_4	Nitrate transport rate 5 cm/d/v Phosphate was not transported	[42]
Lean clay	0.85 V/cm	3.2 g/L NH_4OH 0.48 H_2SO_4	400 mg/kg NH_4OH 200 mg/kg H_2SO_4	[34]
Fine sand	15 $\mu A/cm^2$		Nitrate transported 250 mg/L	[1]
Kaolinite	123 $\mu A/cm^2$		Nitrate transported 250 mg/L	[1]

Table 1. Electrokinetic injections of nutrients.

However, excessive amounts of nutrients in soil exploit the growth and increase the intensity of microorganisms and consequently result in clogging the soil pores causing biofouling [34]. Therefore, it is important to study and carefully plan for the addition of nutrients. A recent study explored the possibility of providing oxygen to polluted soils by electrokinetics for aerobic bioremediation treatments of the soils. Oxygen ions, produced from the electrolysis reaction at the anode, are dissolved in water and transported by electroosmotic flow (see Eq. (1)). Due to the high porosity of the silty and sandy soils, the oxygen ions were transported inside the soil. It was observed that the dissolved oxygen concentrations were high (between 4 and 9 mg/L) in the contaminated zone. The presence of oxygen ions is essential for aerobic biodegradation processes. On the other hand, the transport was not possible in the clay soil.

3.2. Bioaugmentation

Bioaugmentation can be defined as the use of microorganisms which are capable of degrading contamination at specific site. There are two types of bioaugmentation: the first scenario is to the increase the number of bacteria at contaminated site by introducing high number of colony forming units. In this case, the indigenous bacteria is capable of degrading the contaminant, however, the intensity of bacteria is low. In the second scenario, bacterial strain with superior capabilities is introduced to the contaminated site to enhance the degradation process because the indigenous bacterium is not capable of degrading the contaminant. Electroosmotic flow can be used to transport microorganism through contaminated soil to the contaminated zone [11]. For instance, the transport of bacteria in clay and sand by electroosmotic flow and electrophoresis was investigated. The results showed that 20% of bacteria were transported by electrophoresis. A more recent study showed that microorganisms can be transported by electrokinetics in sand via electrophoresis and the microorganisms remained active and viable after the transport process [36]. Another study showed that by adding bacteria in the anode and cathode compartment, bacteria was transported via electroosmotic flow in clay soil. However, in general, bioaugmentation studies have not been successful. The lack of success has been attributed to the formation of antibiotics by indigenous bacteria, predation and adaptability of new bacteria to the contaminated soil. For instance, Pseudomonas sp. LB400 bacteria were found to be capable of degrading 4-chlorobiphenyl in sterilized soil, but a decrease in their viability was observed when non-sterilized soil was used. In many cases, there is a need for the use of more than one bacterial strain to be able to degrade contamination at specific contaminated site. Microbial consortia can degrade contamination with high efficiency, however, bacteria tend to compete with one another. There an urgent need for a new technique that can minimize/eliminate the competition between the bacterial strains. In electrokinetic bioremediation, the application of electric current disrupts bacteria membrane by changing the orientation of membrane lipids. High current can be used to deactivate/kill bacterial strains that have negative impact in the degradation process. Application of high voltages in the range of 25 kV cm^{-1} per 40 μs pulse duration can kill bacteria. The effect of direct current application on different strains of bacteria in liquid and slurries has been investigated. Further research in this area is required to explore the use of electrokinetics as tool to kill the unfavorable bacteria.

Enzymes can be used in bioremediation, instead of microorganism, to avoid the competition between the bacterial strains in the microbial consortia. The new technologies can be used to produce enzymes with high quality and increase the shelf time for the enzymes. Enzymes are simple to use in the bioremediation process and there is no unfavorable produce when using enzymes in the remediation process. The only down side of using enzyme is the high cost associated with the use of enzyme. The delivery of enzymes using electroosmotic flow has not yet been investigated.

4. Electrokinetic processes

Application of electric field in electrokinetic remediation results in electrolysis reactions at the electrodes, electroosmotic flow from anode to cathode, electromigration of ions to the oppositely charged electrode, and electrophoresis. Aforementioned processes can change the physiochemical properties of the soil and subsurface conditions at the contaminated site. Electrolysis reaction results in acidic conditions near the anode and basic environment near the cathode. Application of electric current develops voltage gradients and forms zones with different current intensity. Soil temperature increases due to application of electric current. Contaminants attached to the soil matrix can be desorbed by the effect of electroosmotic flow and contribute to the bioavailability.

4.1. pH gradient

In electrokinetic processes, application of electric current results in occurrence of electrolysis reactions of water (redox) at the electrodes. Oxidation occurs at the positively charged electrode generating hydrogen ions and liberates oxygen gas. Reduction takes at the negatively charged electrode producing hydroxyl ions and hydrogen gas. The hydrogen ions (i.e. H^+) form an acid front that travels toward the cathode by the effect of three mechanisms including: electroosmotic flow, diffusion, and electromigration. The acid front reduces the pH of the soil creating acidic environment. The hydroxide ions form a base front that travels toward the anode by electromigration and diffusion and elevate the pH of the soil in the vicinity of the cathode. The base front is slower than the acid front because the base front moves in a direction that is against the electroosmotic flow and the hydroxyl ions (OH) are heavier in weight than hydrogen ions (H^+). Therefore the acid front and base front meets at a point closer to the cathode. The acidic environment near the anode and alkaline conditions at the vicinity of the cathode play a very important role in the electrokinetic processes. For example, the low pH conditions increase desorption of heavy metals from contaminated site. High and low pH environments have a crucial effect on the survival of bacteria at contaminated site. Heavy metals are soluble at low pH < 6 and precipitate at a pH > 7. In electrokinetic remediation, the soil pH is in around 2–3.5 near the anode and between 8 and 11 at the cathode. For instance, cobalt and copper are tend to dissolve into solutions at pH around 4.5–6.5, and form insoluble hydroxides at pH greater than 7.5. The low pH near the anode contributes positively to the dissolution of heavy metals and enhances the removal process. On the other hand, high pH increases the precipitation of heavy metals in the vicinity of the cathode, and decreases the efficiency of the technique in removing contaminants. In electrokinetic bioremediation, the soil pH plays a dominant role in the success of the process. Bacterial survival and optimum degradation efficiency are directly related to the pH conditions. Bacteria can survive at a pH between 6 and 8. Some bacterial strains can tolerate very high and very low pH values. Bacteria have the capability of controlling the exchange of hydrogen ions through the cell wall to adapt the intracellular pH. However, is has been found that the high pH gradient across cell membrane has a detrimental effect on the growth and metabolism of bacteria.

Researches and engineers have developed many techniques to control pH during electrokinetic remediation; these techniques can be divided into two approaches including conventional and innovative techniques. The conventional techniques include the ion selective membrane technique in which a cation-exchange membrane is introduced to cease the movement of the hydroxide ions from the cathode to the soil as shown in **Figure 2**; continuous changing/removing of the solution in the electrode compartments; and addition of chemical compounds such as ethylenediaminetetraacetic (EDTA), acetic acid, and nitric acid. On the other hand, innovative techniques include a stepwise moving anode, polarity exchange, circulation of an electrolyte (anolyte and catholyte) solution in the electrode compartments (see **Figure 3**), and the two anodes technique (TAT) (see **Figure 4**) which have investigated the control of the advancement of the acid and the base fronts [37]. There are many factors that can

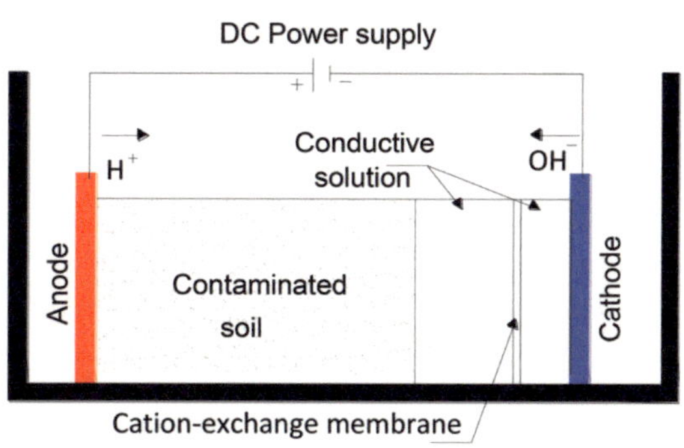

Figure 2. Electrokinetic remediation with ion selective membrane (after [43]).

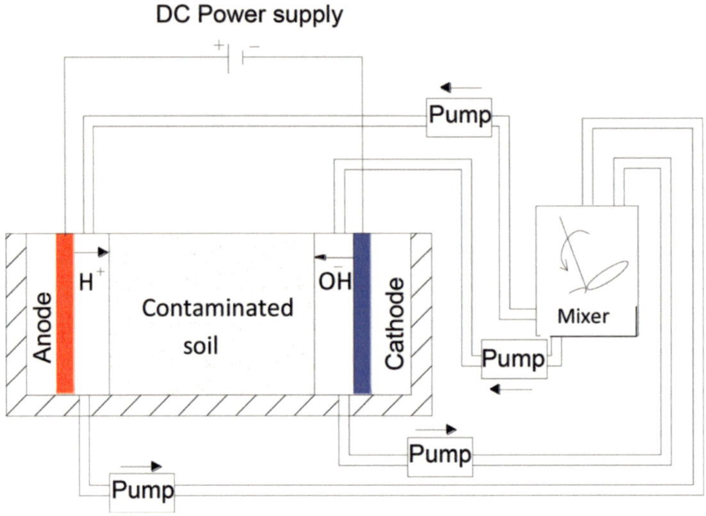

Figure 3. Electrokinetic remediation cell with electrolyte solution circulation technique (after [43]).

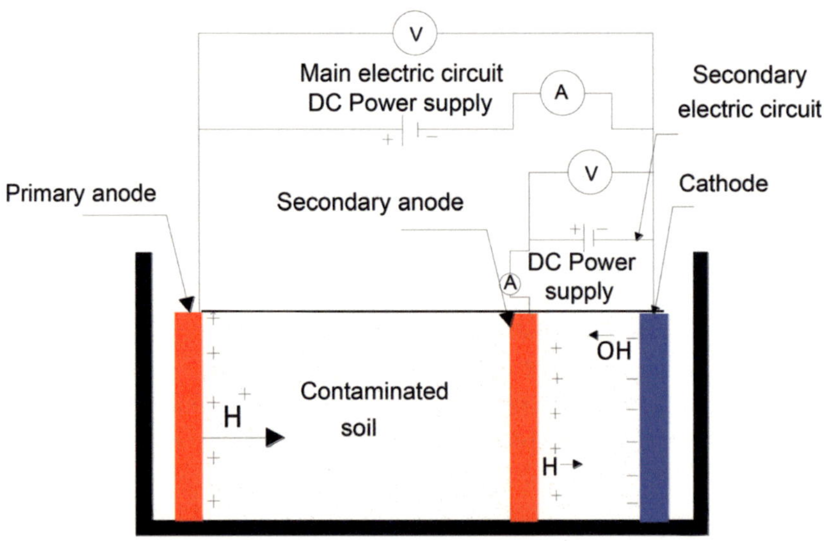

Figure 4. Two anode technique (TAT) (after [43]).

affect the soil pH including the soil type and the soil buffer capacity. The soil puffer capacity is directly influenced by the presence of anions, carbonates, hydrocarbonates, hydroxides, borates, phosphates, silicates, and organic acids anions. These factors should be taken into consideration before selecting the suitable technique to control the soil pH.

Many researchers have investigated the effect of pH on electrokinetic bioremediation using conventional methods. For instance, the use of electrokinetic bioremediation to mitigate creosote-polluted clay soil was investigated. In this study, the soil pH was controlled by continuous substitution of the solutions used at the water compartments. The downside of using this technique is the amount of work involved in the process (removing/changing the solution) and the cost of the process plus this technique is not suitable for field applications. Moreover, in this technique it is required to treat the electrode solution (removed from the electrode compartment) before disposing it. The technique that involves the use of chemical compound is not recommended due to the possibility of generation of chemical reaction that has a negative impact in the remediation process. The addition of acids, such as hydrochloric acid and nitric acid, to reduce the pH near the cathode can result in acidic condition, which it is extremity difficult (if not impossible) to reverse it is effect to the original condition [37]. The innovative techniques, that are available to control pH at contaminated sites, are expensive and in some cases are not applicable in field applications. For example, the step moving anode technique required labor work and involve the mobilization of the anode along the distance between the electrodes. Also, this technique is not suitable for electrokinetic bioremediation as it results in lowering the soil pH and that is not in the favor of the process. This technique is suitable for desorption and mobilization of heavy metals as the moving electrode (anode) generate hydrogen ions which result in lowering the soil pH ($pH \le 4.5$). The low pH environment is suitable for desorption of heavy metals from soil, while it affects bacterial survival in bioremediation processes.

In electrokinetic bioremediation, the low pH condition has adverse effect on indigenous bacteria. The measurement of pH during the polarity exchange technique is crucial for the success of the process. In a research study of phenol-contaminated soil, the polarity reversal technique was used to control the soil pH and water content. This technique can be suitable for electrokinetic bioremediation; however, continuous pH monitoring is required which is challenging and increases the overall cost of the process. In another research study, Kim et al. [34], developed and used a technique involves the circulation of electrolyte solution between the electrode compartments during the process as illustrated in **Figure 2**. This circulation of electrolyte solution can be used to control the pH during electrokinetic bioremediation; however, the issue will be the running cost and the need for maintenance in the field. The circulation of electrolyte solutions can be a challenging and very difficult for implementation in the field application. In electrokinetic bioremediation applications, the control of field conditions, especially pH, is very important for the survival of indigenous bacteria and contributes positively to the success of the process. Many techniques have been developed and implement to control pH during electrokinetics application. However, more research is required to enhance the outcome of the existing techniques. Also, there is a need for development of new techniques to control the pH during electrokinetic bioremediation application.

Recently, the authors of this chapter have investigated an innovative technique that can be used not only to stabilize pH but also to distribute nutrients uniformly inside the contaminated soil during electrokinetic bioremediation [37]. In the innovative technique, two electrodes are placed in each water compartment one serves as an anode and the other as a cathode to form two electric circuits that connected to a power supply. At each compartment, the anode will produce hydrogen ions and the cathode will produce hydroxyl ions, the coexistence of these ions in the same compartment will result in neutralizing each other. As shown in Eqs. (4) and (5), the innovative configuration is supposed to forms the same numbers of hydrogen ions and hydroxide ions with all the ions reacting to form water. The innovative technique provides solutions for challenges facing other techniques; there is no need for continuous pumping or addition of chemical compounds.

4.2. Electric current density and voltage gradient

During electrokinetic bioremediation, microorganisms at contaminated site are subject to stress due to application of electric current. The electric current has direct and/or indirect effects on indigenous microorganisms. For instance, application of high voltage can cause a rupture in the cell membrane. Also, application of electric current can be a companied with chemical reactions that produce by-products which are harmful to the microorganisms. In the food industry, the electric current is used for disinfection purposes (killing bacteria). Studies showed that, the use of DC current causes the death/inactivation of living organisms. Over the last decade, researchers have investigated the influence of electric current on electrokinetic bioremediation treatment.

Microorganisms survival and transport during, electrokinetic remediation, is greatly affected by the application of electric current. **Table 2** presents data from previous studies that investigated the effect of electrical current on the indigenous bacteria. In a research study, the effect of

Medium	Current intensity or voltage gradient used	Highlights/main outcome	Reference
Liquid	20 mA/cm^2	High cell density survive	[43]
Soil (kaolinite)	0.31, 0.63, 1.88, 3.13 mA/cm^2	Optimum current 0.63 mA/cm^2	[43]
Liquid	0.04, 4, 8, 12, 14 mA/cm^2	Optimum electric field density 100 kJ/L	[41]
Liquid	10.2 mA/cm^2	No effect on cell activity	[43]
Glass beads	1.8 mA/cm^2	Low level DC has no effect of cell viability	[43]
Clay and silt	0.314 mA/cm^2	pH changes near the anode is major factor affecting the microbial communities	[8]
Soil	1.0 mA/cm^2	No negative effect on indigenes bacteria	[43]
Hide-soak liquors	2 A	Deactivated bacteria	[3]
Activated sludge	0.5–1.5 mA/cm^2	pH or direct contact caused bacterial inhibition	[43]
Fine-grained soil	2 V/cm	The population of bacteria increased near the cathode	[43]
Sandy loam	0.46 V/cm	Rate of transport is 0.11 cm/h Microorganisms are active after the transport process	[3]
Tap water	0.28–1.4 V/cm	Optimum voltage intensity is between 0.28 and 1.4 V/cm	[8]

Table 2. Effects of electrical current.

electric current on the intensity of indigenous bacteria was investigated. It was concluded that the application of electric current is harmful on the microbial community with low cell densities; however, the electric current was minimal in microbial communities with high cell densities. Recent study, showed that the optimum degradation of an organic compounds (pentadecane in kaolinite soil) that the optimum pollutant removal was achieved using an intermediate electric current density occurs when using electric current intensity of 0.63 mA/cm^2 compared with the higher and lower current densities of 3.13 and 1.88 mA/cm^2, respectively. Another study showed that selecting electric current intensity is essential not only in obtaining optimum degradation but also in retaining the indigenous microorganisms. This study showed that 37% of total petroleum hydrocarbons were degraded in the vicinity of the anode with an optimum electric field of 2 V/cm.

Previous studies showed that the use of electroosmotic and electromigration (electric field) to transport microorganisms did not reduce the capability of microorganism in degrading organic matter [36]. Very few studies investigated the effect of the electrode materials on the electrokinetic bioremediation. For instance, the results of an experimental study showed that indigenous microbial community is adversely affected by the products of the electrochemical reactions between the electrode material and the soil medium. The chemical reactions between the electrode material and the soil medium depend on many factors and it is very difficult to predict the by-products. It has been observed that not only the electric current intensity affects the microbial survival, but more importantly the combined effect of applied current intensity and duration is the crucial factor affecting living organisms. The use of steel, copper, and carbon as electrodes with different combination (anode-cathode) in electrokinetic remediation was investigated. The results showed that the efficiency of the remediation process is significantly affected by the selection of the electrode material and which material to be used in the anode and in the cathode. More research is needed in this area to investigate the effect of electrode materials in electrokinetic bioremediation.

4.3. Temperature

It is known that microorganisms can survive in various environmental conditions including a wide range of temperature. For instance, thermophile can tolerate temperature between 45 and 120°C, mesophile can live in temperature between 20 and 45°C, and psychrophile survive at low temperatures between −20 and 10°C. In general, microorganisms' growth rate is directly proportional to the temperature between 25 and 34°C, also the increase in temperature results in increase of metabolism and the highest degradation occurs at temperature between 27 and 42°C. The increase in temperature during electrokinetic processes is reported in the available literature. For example, a recent study showed that application of electric current resulted in an increase in the soil temperature between 5 and 20°C, the maximum increase was observed in the vicinity of the anode. Another study showed that the temperature increased up to 90°C during field application and intermittent current was used to interrupt the electric current so as to reduce the temperature. Although, the increase in temperature during electrokinetic remediation is known and well documented, yet, very few reports in the current literature discussed the effect of temperature on electrokinetic bioremediation. Researchers tend to attribute the increase in biodegradation to nutrient delivery by electrokinetic and over looked other factors that contribute positively to the process (i.e. temperature). In the current literature there are very few reports that studied the implication temperature in the electrokinetic bioremediation. In a previous study, electrokinetic was used to deliver nutrients and oxygen to microorganisms in the contaminated zone. It was concluded that the increase in soil temperature, resulted from application of electric current, enhanced the degradation process. The continuous application of electric current using high applied voltage for long period of time can increase the temperature inside the soil being treated. No doubt, the elevated temperature has a negative impact on the viability of indigenous bacteria. Intermittent current can be used to eliminate the effect of continuous current in increasing the soil temperature. The use of current intermittence can control the increase in temperature, and also can contribute positively to the outcome of the remediation process.

4.4. Bioavailability

Bioavailability can be measured by the amount of colony forming unit (CFU) of bacterial strains that capable of degrading the contaminants, the CFU must be available in soil liquid (pore fluid) at a given time. Bioavailability is also can be defined as the portion of contaminants that is available (ready) to be consumed by microorganisms without the need for desorption process (**Table 3**). When the contaminant come in contact with the soil, sorption of the pollutants by the soil will take place and the sorption rate is depending on environmental conditions including pH, temperature, etc. After sorption took place, the bioavailability of the contaminant will mainly be dependent on the back-diffusion process. Therefore, back-diffusion plays a dominant role in controlling the bioavailability of contaminants. The main two schools of thought concerning bioavailability are: (1) bacteria can degrade a contaminant regardless to the sorption and desorption process, even if it is attached to the soil matrix; (2) desorption of contaminants from soil is a prerequisite for the degradation process to occur (desorption of pollutants should take place first before microorganisms can degrade it). In electrokinetic remediation, electroosmotic flow creates flow net to the soil solids (within the double layer) therefore, electroosmotic flow can promote desorption of contaminants from soil matrix. In a previous study, the authors have compared the efficacy of electroosmotic flow and hydraulic

Medium	Contaminant	Contaminant concentration (mg/kg)	Highlights/Main outcome	Reference
Fine soil (from a contaminated site)	Petroleum hydrocarbon	78,600	37% reduction	[43]
Sand	Diesel	6800	60% reduction	[35]
Clayey soil	Phenanthrene	200	65% removal	[35]
Coarse sand/sand	Creosote	50, 200, 500, 900, 6800	50, 68, 80% reduction	[8]
Sandy loam	Phenol	180	58% reduction	[8]
Clay	Creosote	1300	35% reduction	[43]
Kaolinite	Pentadecane	1000, 5000, 10,000, 20,000	77.6% reduction	[34]

Table 3. Laboratory tests using different contaminants.

flow in promoting desorption of phenanthrene from clay soil [38]. The results showed that the concentration of phenanthrene in the effluents from the test conducted using electroosmotic flow are three to four times higher than the concentration in desorption test using hydraulic flow. Also, it was found that the energy consumed during the hydraulic flow tests was three orders of magnitude higher than the energy used during the electroosmotic flow tests.

4.5. Available power sources for electrokinetics

Electrokinetic remediation requires electric power to apply voltage gradient between the electrodes, the energy consumption is consider as the factor number one that contribute to the total cost of the process. Therefore, the increase in energy cost results in an increase of the overall cost of the remediation process. No doubt the cost is one of the major factors affecting the selection of suitable remediation technique, thus high cost of a remediation process can be a major obstacle restricting wide field applications of this technology. Previous studies showed that the cost of energy represents 30% of the total cost of an electrokinetic remediation process; however, very few studies have discussed an alternative economical source of power that can reduce the cost of energy. Power generated by solar panels is renewable energy and has no negative impact on the environment.

Recently, the environmental awareness increases and the solar energy has become the center for the interest of scientists and the public as environmentally friendly source of power. Recent report from Solar Buzz showed that more than 70% of the photovoltaic (PV) resources have been installed in northern hemisphere including countries such as Germany, Japan, USA, and Canada. Previous studies showed that the efficiency of the solar panel increased during the winter because of the cold weather. Although, power generated by solar panels can be an excellent candidate for power supply in electrokinetics, yet there are very few reports that have investigated the use of solar power as a source of power in electrokinetic bioremediation. The effect the off power period, during the night, on the remediation process is not presented in the literature. Some of the advantages of using solar panels to generate power for electrokinetic remediation are the elimination of the cost of electricity transmission and the reduction of power losses in the transmission lines. Solar panels produce direct current (DC) field that can be used in electrokinetic remediation without the need for a transformer. In the near future, it is expected that the solar cell prices will decrease and as the technology continues to improve that can significantly increase the efficiency of the solar panels. The power generated by solar panel fluctuates during the day (starts from zero before sun rise and increases until noon time and decreases to zero by the sun set) and it is directly dependent the weather conditions (sunny, cloudy, rainy, etc.). This can cause disruption in the power supply during the day and intervals of zero voltage at night, especially in the northern latitudes with little day light during winter.

The application electric field during electrokinetic remediation results in ions orientation in the pore fluid that resists electric current. The ions orientation reduce the efficiency of the remediation process, however, the interruption of the electric field allows the restoration of original ions orientation, which can enhance the remediation process. The non-stabilized electric current generated by solar panels can stimulate the remediation process by the effect of restoration of ions orientation as explained earlier. Many studies have proven that the use of current intermittence can enhanced the outcome of an electrokinetic remediation. In a previous study, the authors have investigated the use of solar panels to generate power for the electrokinetic remediation to remediate soil contaminated with copper [39]. Three solar panels were used to generate electric field across the contaminated soil 41, 27 and 13.5 V. The results showed that the power generated by the solar panels was enough for mitigation of soil contaminated with heavy metals. In recent work, the authors used solar panels to generate power for the electrokinetic bioremediation to mitigate clay soil contaminated with phenanthrene. The results showed that solar panels can be used successfully to produce enough power for electrokinetic bioremediation of petroleum hydrocarbons. Moreover, the intervals of zero voltage at night can decrease soil temperature in field applications, which is a benefit.

5. Electrokinetic-enhanced phytoremediation of contaminated soils

Pollutants are released to the environment from anthropogenic activities and natural weathering processes. Human activities such as mining, industrial, and agricultural activities typically lead to contamination of soils by heavy metals and organic pollutants. Some heavy metals and PAH compounds have been identified as mutagenic, carcinogenic, and/or teratogenic contaminants.

Phytoremediation is a green and sustainable process in which plant species are used to remove, degrade, or sequester pollutants from soil or groundwater. Phytoremediation has been shown to be efficient and economically feasible for the treatment of large areas with low contaminant concentrations as compared to other remediation treatments. Both organic and inorganic contaminants can be removed or degraded by the growing plants by several mechanisms, namely; phytodegradation, rhizofiltration, phytoaccumulation, rhizodegradation, and phytostabilization. Phytodegradation is the breakdown of the absorbed organic chemicals by plant metabolic processes or by compounds produced by the plant. Phytodegradation of petroleum hydrocarbons was achieved in a soil that was also contaminated with heavy metals. Rhizofiltration is the removal of contaminants such as heavy metals and radionuclides in the soil by the root membranes [40]. Phytoaccumulation is the incorporation of inorganic chemicals such as heavy metals in plant tissues. Rhizodegradation is the degradation of organic contaminants near the root mass by bacteria and fungi stimulated by the root exudes and enzymes released by the plant. Recent study showed that phytoremediation with *Mirabilis Jalapa* L. removed 63% of total petroleum hydrocarbon from a contaminated soil in 127 days. Another study showed that *Medicago sativa* Linn and Fire Phoenix had removed 80–100% of eight PAH compounds from contaminated soils in 150 days. Phytostabilization is achieved by induced chemical changes to specific heavy metal contaminants causing the metal to become less bioavailable. For example, deep rooting plants could reduce the highly toxic Cr(VI) to Cr(III), which is much less soluble, that is, less bioavailable.

In the decade, the combination of phytoremediation and electrokinetic remediation has been proposed in an attempt to avoid the limitations of phytoremediation. Electrokinetic-enhanced phytoremediation technology consists of the application of a low intensity electric field to the contaminated soil in the close vicinity of growing plants. In this hybrid technology, the removal or degradation of the contaminants is performed by the plant, whereas the electric field enhances the plant activity by increasing the bioavailability of the contaminants by desorption and transport of the contaminants. In recent research studies, electrokinetic phytoremediation technology with electric field applied between 8 and 24 h/day has shown promising results for the restoration of heavy metal contaminated soils. The use of electrokinetic-assisted phytoremediation for soils with hydrocarbon contaminants has not been reported in the literature.

6. Conclusion

Electrokinetic remediation has received more attention in recent years as a well-developed technique for in situ soil decontamination. The principles of electrokinetics have been recently explored to enhance bioremediation and phytoremediation treatment of contaminated sites. Electrokinetics requires low-level electric current which makes solar energy by lower end solar panels an excellent source of power for the technology. Solar energy is an innovative power alternative for electrokinetic-enhanced bioremediation/phytoremediation and can be economically viable, in particular for remote sites with no active power lines.

The electrokinetic bioremediation is a promising technique that can be used to mitigate contamination with organic and inorganic compounds. The main challenges facing the application of electrokinetic are the process cost and the pH gradient. The application of electric current results in high energy consumption that increases the overall cost of the remediation process. This is considered one of the major factors restricting the field application of the technology. Very few studies in the current literature have addressed the cost of energy in electrokinetic bioremediation. Research to date has shown that electrokinetic bioremediation can be used to mitigate contaminated soil, however, the results showed a low to moderate percentage of contaminant degradation. Future research is needed to address the optimization and the removal efficiency of the electrokinetic bioremediation. Researchers and engineers address the pH issue using two different approaches conventional and innovative techniques. The conventional techniques involve the addition of chemical compounds to control the pH. The innovative techniques involve circulation/mixing of the electrolyte solution at the anode and cathode compartment to neutralize the pH. Both conventional and innovative techniques can result in a further increase in the overall cost of remediation process. Moreover, the effect of the increase in temperature associated with electrokinetic bioremediation has not been fully investigated.

The recent advancement in biotechnology and bioengineering provides incredible opportunities for enhancing electrokinetic bioremediation and phytoremediation. The application of electrokinetic bioremediation/phytoremediation involves various microorganisms aspects including the challenges facing the microbes survival in contaminated sites; isolation and characterization of bacterial strains with superior degradation capabilities in bioremediation; the improvement of the metabolism of microorganisms through manipulation using genetic engineering; identifying powerful degrading enzymes using metagenomics; and exploring new engineering ways to enhance bioremediation.

It is anticipated that, with intense research effort, the promising potential of the hybrid electrokinetic bioremediation/phytoremediation technology will encourage further research in order to optimize the efficiency and expedite the transfer of the technology for in situ applications. Carefully designed, simulated and monitored laboratory and field studies are necessary to explore the fundamental and practical aspects of the technology. It is important to simulate and test different electrode configurations and identify hyperaccumulator plants of heavy metals and PAH compounds. It is also necessary to understand the influence of electric field in the physiology of the plant and soil microflora as well as the geobiochemistry involved in the degradation of organics in the rhizosphere.

Author details

Ikrema Hassan[1], Eltayeb Mohamedelhassan[2], Ernest K. Yanful[1] and Ze-Chun Yuan[3*]

*Address all correspondence to: zyuan27@uwo.ca

1 University of Western Ontario, London, Canada

2 Department of Civil Engineering, Lakehead University, Thunder Bay, Canada

3 London Research and Development Centre, Agriculture and Agri-Food Canada, London, Canada

References

[1] Acar YB, Rabbi MF, Ozsu EE. Electrokinetic injection of ammonium and sulfate ions into sand and kaolinite beds. Journal of Geotechnical and Geoenvironmental Engineering. 1997;**123**(3):239-249. DOI: 10.1061/(ASCE)1090-0241

[2] Chen M, Xu P, Zeng G, Yang C, Huang D, Zhang J. Bioremediation of soils contaminated with polycyclic aromatic hydrocarbons, petroleum, pesticides, chlorophenols and heavy metals by composting. Applications, microbes and future research needs. Biotechnology Advances. 2015;**33**(6 pt 1):745-755. DOI: 10.1016/j.biotechadv.2015.05.003

[3] Yeung AT, Gu YY. A review on techniques to enhance electrochemical remediation of contaminated soils. Journal of Hazardous Materials. 2015;**195**:11-29. DOI: 10.1016/j.jhazmat.2011.08.047

[4] Acar YB, Alshawabkeh AN. Principles of electrokinetic remediation. Environmental Science & Technology. 1993;**27**(13):2638-2647. DOI: 10.1021/es00049a002

[5] Reed BE, Berg MT, Thompson JC, Hatfield JH. Chemical conditioning of electrode reservoirs during electrokinetic soil flushing of Pb-contaminated silt loam. Journal of Environmental Engineering—ASCE. 1995;**121**(11):805-815. DOI: 10.1061/(ASCE)0733-9372(1995)121:11(805)

[6] Yeung AT, Corapcioglu M. Electrokinetic flow processes in porous media and their applications. Advances in Porous Media. 1994;**2**:309-395

[7] Alshawabkeh AN. Electrokinetic soil semediation: Challenges and opportunities. Separation Science and Technology. 2009;**44**(10):2171-2187. DOI: 10.1080/01496390902976681

[8] Reddy KR, Cameselle C. Electrochemical Remediation Technologies for Polluted Soils, Sediments and Groundwater. Hoboken, NJ: Wiley; 2009. 732 p. ISBN: 978-0-470-38343-8

[9] Chilingar GV, Loo WW, Khilyuk LF, Katz SA. Electrobioremediation of soils contaminated with hydrocarbons and metals: Progress report. Energy Sources. 1997;**19**(2):129-146. ISBN: 978-0-471-73778-0

[10] Shapiro AP, Probstein RF. Removal of contaminants from saturated clay by electroosmosis. Environmental Science & Technology. 1993;**27**(2):283-291. DOI: 10.1021/es00039a007

[11] DeFlaun MF, Condee CW. Electrokinetic transport of bacteria. Journal of Hazardous Materials. 1997;**55**(1–3):263-277. DOI: 10.1016/S0304-3894(97)00023-X

[12] Nyer EK, Suarez G. Treatment technology: In situ biodegradation is better than monitored natural attenuation. Ground Water Monitoring and Remediation. 2002;**22**(1):30-39. DOI: 10.1111/j.1745-6592.2002.tb00651.x

[13] Luo QS, Zhang XH, Wang H, Qian Y. The use of non-uniform electrokinetics to enhance in situ bioremediation of phenol-contaminated soil. Journal of Hazardous Materials. 2005; **121**(1–3):187-194. DOI: 10.1016/j.jhazmat.2005.02.007

[14] Inglis TJJ, Sagripanti JL. Environmental factors that affect the survival and persistence of *Burkholderia pseudomallei*. Applied and Environmental Microbiology. 2006;**72**(11):6865-6875. DOI: 10.1128/AEM.01036-06

[15] Locatelli A, Spor A, Jolivet C, Piveteau P, Hartmann A. Biotic and abiotic soil properties influence survival of *Listeria monocytogenes* in soil. PLoS One. 2013;**8**(10):e75969. DOI: 10.1371/journal.pone.0075969

[16] Flemming H-C, Wingender J, Szewzyk U. Biofilm Highlights. Dordrecht: Springer; 2011. 243 p. DOI: 10.1007/978-3-642-19940-0

[17] Malik A, Grohmann E, SpringerLink. Environmental Protection Strategies for Sustainable Development. Dordrecht: Springer; 2012. 591 p. DOI: 10.107/978-94-007-1591-2_1

[18] Maldonado LA, Quintana ET. Unexpected properties of micromonosporae from marine origin. Advances in Microbiology. 2015;**5**:452-456. DOI: 10.4236/aim.2015.56046

[19] McKenney PT, Driks A, Eichenberger P. The *Bacillus subtilis* endospore: Assembly and functions of the multilayered coat. Nature Reviews Microbiology. 2013;**11**(1):33-44. DOI: 10.1038/nrmicro2921

[20] Abel-Santos E. Bacterial Spores: Current Research and Applications. Norfolk, UK: Caister Academic Press; 2012. DOI: 10.1086/675034

[21] Qin JF et al. Modular pathway rewiring of *Saccharomyces cerevisiae* enables high-level production of L-ornithine. Nature Communications. 2015;**6**:8224. DOI: 10.1038/ncomms9224

[22] Sah S, Aluri S, Rex K, Varshney U. One-carbon metabolic pathway rewiring in *Escherichia coli* reveals an evolutionary advantage of 10-formyltetrahydrofolate synthetase (Fhs) in survival under hypoxia. Journal of Bacteriology. 2015;**197**(4):717-726. DOI: 10.1128/JB.02365-14

[23] Kang A, Chang MW. Identification and reconstitution of genetic regulatory networks for improved microbial tolerance to isooctane. Molecular Biosystems. 2012;**8**(4):1350-1358. DOI: 10.1039/C2MB05441H

[24] McErlean C, Marchant R, Banat IM. An evaluation of soil colonisation potential of selected fungi and their production of ligninolytic enzymes for use in soil bioremediation applications. Antonie Van Leeuwenhoek International Journal of General and Molecular Microbiology. 2006;**90**(2):147-158. DOI: 10.1007/s10482-006-9069-7

[25] Karigar CS, Rao SS. Role of microbial enzymes in the bioremediation of pollutants: A review. Enzyme Research. 2011;(2001):1-11. DOI: 10.4061/2011/805187

[26] Brenner K, You LC, Arnold FH. Engineering microbial consortia: A new frontier in synthetic biology. Trends in Biotechnology. 2008;**26**(9):483-489. DOI: 10.1016/j.tibtech.2008.05.004

[27] Schneewind O, Missiakas DM. Protein secretion and surface display in Gram-positive bacteria. Philosophical Transactions of the Royal Society B-Biological Sciences. 2012; **367**(1592):1123-1139. DOI: 10.1098/rstb.2011.0210

[28] Ray D, Mohandass SN, Roy P. Molecular phylogeny of arctic microbes using metagenomic approach. Advances in Microbiology. 2014;**4**:1278-1284. DOI: 10.4236/aim.2014.416137

[29] Yergeau E, Sanschagrin S, Beaumier D, Greer CW. Metagenomic analysis of the bioremediation of diesel-contaminated Canadian high actic soils. PLoS One. 2012;**7**(1):e30058. DOI: 10.1371/journal.pone.0030058

[30] Miller MB, Bassler BL. Quorum sensing in bacteria. Annual Review of Microbiology. 2001;**55**:165-199. DOI: 10.1146/annurev.micro.55.1.165

[31] Rutherford ST, Bassler BL. Bacterial quorum sensing: Its role in virulence and possibilities for its control. Cold Spring Harbor Perspectives in Medicine. 2012;**2**(11):pii (1-25)a012427. DOI: 10.1101/cshperspect.a012427

[32] Huang YL, Zeng YH, Yu ZL, Zhang J, Feng H, Lin XC. In silico and experimental methods revealed highly diverse bacteria with quorum sensing and aromatics biodegradation systems—A potential broad application on bioremediation. Bioresource Technology. 2013;**148**:311-316. DOI: 10.1016/j.biortech.2013.08.155

[33] Madsen EL. Determining in situ biodegradation—Facts and challenges. Environmental Science & Technology. 1991;**25**(10):1663-1673. DOI: 10.1021/es00022a001

[34] Kim SJ, Park JY, Lee YJ, Lee JY, Yang JW. Application of a new electrolyte circulation method for the ex situ electrokinetic bioremediation of a laboratory-prepared pentadecane contaminated kaolinite. Journal of Hazardous Materials. 2005;**118**(1–3):171-176. DOI: 10.1016/j.jhazmat.2004.10.016

[35] Xu W, Wang CP, Liu HB, Zhang ZY, Sun HW. A laboratory feasibility study on a new electrokinetic nutrient injection pattern and bioremediation of phenanthrene in a clayey soil. Journal of Hazardous Materials. 2010;**184**(1–3):798-804. DOI: 10.1016/j.jhazmat.2010.08.111

[36] Maillacheruvu KY, Chinchoud PR. Electrokinetic transport of aerobic microorganisms under low-strength electric fields. Journal of Environmental Science and Health Part A—Toxic/Hazardous Substances & Environmental Engineering. 2011;**46**(6):589-595. DOI: 10.1080/10934529.2011.562832

[37] Hassan I, Mohamedelhassan E, Yanful EK, Bo MW. A new technique for enhancing electrokinetic bioremediation. Environmental Geotechnics. 2016. E-ISSN 2051-803X. DOI: 10.1680/jenge.16.00001

[38] Hassan I, Mohamedelhassan E, Yanful EK, Yuan Z-C. Sorption of phenanthrene by kaolin and efficacy of hydraulic versus electroosmotic flow to stimulate desorption. Journal of Environmental Chemical Engineering. 2015;**3**(4, Part A):2301-2310. DOI: 10.1016/j.jece.2015.08.011

[39] Hassan I, Mohamedelhassan E, Yanful EK. Solar powered electrokinetic remediation of Cu polluted soil using a novel anode configuration. Electrochimica Acta. 2015b;**181**:58-67. DOI: 10.1016/j.electacta.2015.02.216

[40] Lee M, Yang M. Rhizofiltration using sunflower (*Helianthus annuus* L.) and bean (*Phaseolus vulgaris* L. var. *vulgaris*) to remediate uranium contaminated groundwater. Journal of Hazardous Materials. 2010;**173**(1–3):589-596. DOI: 10.1016/j.jhazmat.2009.08.127

[41] Tiehm A, Augenstein T, Ilieva D, Schell H, Weidlich C, Mangold KM. Bio-electro-remediation: Electrokinetic transport of nitrate in a flow-through system for enhanced toluene biodegradation. Journal of Applied Electrochemistry. 2010;**40**(6):1263-1268. DOI: 10.1007/s10800-010-0101-2

[42] Schmidt CAB, Barbosa MC, de Almeida MDS. A laboratory feasibility study on electrokinetic injection of nutrients on an organic, tropical, clayey soil. Journal of Hazardous Materials. 2007;**143**(3):655-661. DOI: 10.1016/j.jhazmat.2007.01.009

[43] Hassan I, Mohamedelhassan E, Yanful EK, Yuan Z-C. A review article: Electrokinetic bioremediation current knowledge and new prospects. Advances in Microbiology. 2016; **6**(1):57-75. DOI: 10.4236/aim.2016.61006